Code School

Overcome Imposter Syndrome.
Kick-Start Your Tech Career.

MORGAN LOPES AND TIM WHITACRE

Lead Editor: Emily Buchanan
Editorial Support: Steph Whitacre
Cover Design by Kaci Ariza
Interior Layout by Darren Kizer
Typeset in Adobe Garamond Pro and Barlow Condensed

For more information, email mail@codeschoolbook.com.

ISBN: 978-1-7360807-0-2

KICK-START YOUR CAREER ONLINE

This book is just the beginning. We have also put together a handful of free online resources to help you on your journey.

For starters, make sure to visit codeschoolbook.com for more about the book, bios of our contributors, and curated resources mentioned throughout the book.

We also publish a newsletter at codeschoolnewsletter.com. Sign up to get free resources and job postings in your inbox.

Our goal is to help in any way we can. We've done our best to provide a good start here, but if there is something we missed, feel free to reach out at mail@codeschoolbook.com.

Go boldly forward.

DEDICATION

To Jason Ardell, my mentor and friend. You changed my life. Thanks for deeming it work worth doing. – Morgan

To my former students. Your dedication was inspiring. Your perseverance paid off. I raise an Iron Pint to you. – Tim

TABLE OF CONTENTS

PREFACE

I (Morgan) was a high school senior embarking on spring break. Instead of parties at the beach, I opted for a trek into the Appalachian Mountains with two friends. Known by locals as the AT, the Appalachian Trail is a 2,200-mile path stretching from Springer Mountain in Georgia to Mount Katahdin in Maine. Our goal was to spend a week in the woods, hiking 15 miles per day, sleeping on the ground, and carrying everything we needed in 50-pound packs. This was far from the average American Spring Break.

For nearly 100 years, the AT has been a well-worn retreat for locals and visitors from around the world. Regularly maintained paths, sturdy shelters, and pristine markings highlight the way for travelers of all ages. Families enjoy segments of the trail for afternoon walks. Adventurers spend months hiking from end to end. Both weekend warriors and weathered wanderers find the trail approachable. Approachable enough, in fact, for three high schoolers to convince their cautious parents to let them venture into the wilderness. We experienced the allure firsthand as the trail greeted us warmly.

As we hiked, the experience made an impression. Even as a teenager, I noticed the thoughtful maintenance of the trail. Massive segments are kept up by volunteers. Guardians regularly groom brush, remove debris, and trim overgrowth each season. From time to time, we came across strangers handing out water and supplies on the roadside. Hiking companies arranged pickups and dropoffs at outposts to keep hikers equipped and fueled. There was even a comradery and connection among disparate travelers on the journey.

Everything was marked. White blazes indicated the path. Black diamonds were painted on trees, indicating shelter every dozen miles. Blue squares marked fresh water sources, whether immediately obvious or miles off the path. There were placards for historical landmarks and signage positioned at key points along the way. I later came to recognize this as wayfinding.

Wayfinding helps travelers navigate complex environments. It's the byproduct of thoughtful investments from former explorers, historians, and guides. The assistance enhances the understanding and experience for novices and newcomers, making their surroundings more approachable. The greater the complexity, the more valuable the wayfinding. From generation to generation, wayfinding equips each new participant to benefit from the collective knowledge of those who came before them.

In preparation for our trip, my friends and I read books, bought maps, and sought expertise. The resources, specifically for the Appalachian Trail alone, seemed endless to us at the time. In sharing our journey with others, we realized friends and family we'd known for years had their own stories to share about the trail. In the days before our trip, one book came up over and over again: *A Walk in the Woods* by Bill Bryson.[1] In it, Bryson described his own experience of hiking the AT. His ability to share both the joys and pitfalls he found along the trail have encouraged many aspiring hikers throughout the years. *A Walk in the Woods* was, and still is, the de facto resource for people hiking the Appalachian Trail.

No such book exists for individuals transitioning into technology. For those choosing code school, the support is particularly bare. The need for better wayfinding along this unique path brought me here to write one. I've written hundreds of software applications. I've mentored dozens of technology professionals. I've coached and led alongside teams of all sizes. Themes emerged throughout, with the most prominent being this: New professionals arrive eager and ambitious, only to find the infrastructure lacking. Time and energy are wasted as people wander around, hoping for a clear path forward.

Unfortunately, many burn out before they find their way.

Technology has neglected wayfinding for early stage professionals to get oriented. The trails are not clearly marked and pitfalls remain hidden.

In many ways, entering the tech industry is akin to being dropped in a densely wooded forest. Imagine a thick canopy obscuring the sunlight. Brush and grass stretch above your knees. Your shoes immediately sink in the mud. There is no one around nor a discernable path. It is disorienting, discouraging, and debilitating. Without help in sight, at best you may fumble your way along, at worst you may begin to panic.

If the next step feels unclear, that's because it hasn't been marked. Fortunately, you are not as lost as it might appear. Millions before you have felt the same fear and uncertainty. You are not alone. For years, I have been walking the journey with friends, mentors, and peers, many of whom contributed to this book. Tim Whitacre joined me in writing it, and as you read you will be hearing experiences and insights from our own journeys. Both of us managed to find our way through the industry without a computer science degree. Joined by friends and colleagues, we are passionate about helping others discover their right next step.

When I think back to my spring break on the AT, I recall the serendipity that often befell us. At times, it was almost eerie. They called it "trail magic" when out of nowhere a sporadic, delightful event would lift your spirits. One such moment occurred during the first night of our trip. Fellow campers invited us to enjoy fresh brisket chili around their campfire. The next afternoon, following the day's most exhausting terrain, we found an entire package of sugar-coated, peach flavored gummies sprawled across the trail. Hours later, we discovered a beautiful spread of refreshments at a random road crossing. Even at our lowest moments, which were many, inspiration was never too far behind.

As you read *Code School*, I hope you find direction and come across a bit of trail magic of your own. We have marked out important portions of the path. Like a map in your backpack, let this book serve as your Wayfinder.

Welcome to the trailhead.

Mindset

ENGINEERS SOLVE PROBLEMS
Excellence is a byproduct of diligence and effort, not a requirement to get started.

Hacker, coder, developer, programmer… How many names and titles are floating out there for software engineers? Each one is trying to convey the same concept: these people are able to talk to computers. For most terms, however, their definition is limited. A chef at the grill isn't merely a burger-flipper. A pianist composes more than noisy keystrokes. A surgeon's ability goes far beyond scalpel wielding. Most titles overemphasize a specific skill and misrepresent the true work of a technologist. Our value exceeds the ability to merely write code, and our work is more craft than skill. Internally, we know this—but do we know why it matters? The language we use to describe ourselves influences how we perceive ourselves and the work we do. When we trivialize our work, others will too.

I'm reminded of a time in college. I noticed a friend of mine checking her blood sugar. I leaned over and said, "I didn't realize you were diabetic." Almost immediately she replied, "I'm not." Confused, I asked, "Then why are you checking your blood sugar?" She said, "I do have diabetes but I am not a diabetic. Diabetes doesn't define me."

It took me a few seconds to understand her message. What seemed a semantic difference to me was a question of identity to her. Our words have power. As I entered programming, this learning stuck with me. It's why I prefer the term "software engineer." Coders code. Programmers write programs. Engineers solve problems. The distinction feels important.

The concept of engineering has a deep legacy. Original engineers used mathematics to design, construct, and operate military siege engines. The term has seen multiple evolutions but you'll consistently discover meanings that imply specialist, craftsman, and architect. Engineers combine information, resources, and expertise to manage tradeoffs and craft the optimal solution. Code is one of many tools within an engineer's repertoire. Writing code is part of the equation, not the main point.

Reflecting on what makes a great engineer, a few statements come to mind:

- Engineers produce thoughtful, clean, and organized work.
- Engineers write as little code as necessary.
- Engineers consider second and third order effects.
- Engineers work to simplify complex systems.
- Engineers learn consistently.
- Engineers solve problems.

It might seem early to discuss the mindset of engineering and problem solving, as these behaviors seem most relevant to life on the job. But we raise the point now because signs of a strong engineering mindset can be seen in the early days of code school. In many cases, a hunger to solve problems predates programming altogether.

How do you respond to challenges? Which life circumstances have you overcome, instead of allowing them to become excuses? How do you eliminate the countless barriers that interfere with what you want? In what ways are you working around life's roadblocks and setbacks to pursue a new career?

Friend and self-taught engineer, Brandon McLean says, "Coding seems so daunting to some. They would rather draw inside the lines because they are afraid to go against guides, tutorials or what's taught. Sometimes you have to try things and fail to properly learn some aspects of coding." You've got to have grit.

I have watched students drive three hours a day, manage two jobs, and maintain family obligations to complete code school. Some juggled it all at once. It was inspiring. They benefited from the community and lessons, but the most valuable preparation came from overcoming obstacles in their own lives to participate. Tackling these challenges were not simple inconveniences but examples of the tenacity which underpins the most important skill of their careers: mindset.

The need for tenacity and persistence never ends. Terrence Jackson, a fellow software engineer and tech entrepreneur, captures an important point, "Seasoned engineers don't always know what they are doing. Despite popular opinion, highly successful people generally are not the smartest people in the room. Instead, it's the engineer who can formulate a thoughtful approach and arrive at a functional solution." Through exposure to the limits of our understanding, we test our knowledge and learn new things.

Jackson continues with an important point about humility and asking for help, "If you run into obstacles, that is natural. Stick with it and don't be afraid to ask for help. The best way to learn is to keep trying. Do not confuse overwhelm with defeat." While engineers constantly seek answers, they are imperfect and will sometimes get it wrong. Throughout my research for this book, themes like this came up over and over again. Excellence is a byproduct of diligence and effort, not a requirement to get started.

Ryan Holiday, an American author and marketer, wrote a helpful book, *The Obstacle is the Way*, that speaks to this point.[1] In the book, Holiday consolidates ancient philosophies about struggle and adversity. As the title suggests, the setbacks and hardships we endure are part of the journey. Rather than viewing obstacles as merely something to overcome or a cost of our ignorance, these barriers are tools to test our resolve. More valuable than the destination is the overcoming we accomplish in the process.

Many face setbacks and conclude, "It must not be meant to be." Others encounter the same problems and ask themselves, "If this is the cost, how much do I want it? What would I sacrifice to get it?" The situation is the same, but the outcome is determined by perspective.

Without problems to solve, there would be no need for engineers. The discipline and work required are not for everyone. But for those who enjoy creating solutions and persevering through challenges, there are endless opportunities to exercise your interests. There is an ocean of fascinating puzzles and problems, available at a relatively low cost. It's my favorite part of software engineering. With few tools, you unlock an entire world of creativity.

Your first step may be to determine if you're the right person for this style of work. If not, how might you become such a person? It's your choice. Do not let the speculation and judgement of others dissuade you. All knowledge within tech can be acquired, especially with an understanding of how an engineer works.

KNOWLEDGE WORK
In technology, the outward appearance fails to represent what's happening within the minds and imaginations of those in front of the screen.

One winter, I traveled to Colorado for a snowboarding trip. Feet of fresh snow had fallen in a matter of days and the weather forecast looked beautiful. We met up with friends who lived minutes from the ski slopes. We were eager for a few days in the mountains.

In Colorado, one of my friends had already spent over 60 days on the slopes that season. He knew each route like the back of his hand. While we were gearing up, he turned to me and asked, "Where is your helmet?" Given my experiences snowboarding back home, I hadn't even considered it.

Before this trip, my snowboarding experience was limited to smaller mountains on the East Coast. There, the snow is mostly fake, machine blown ice and the runs are a fraction of the length. Either due to the beach-like weather or carnivalesque lifts, snow sports in the southeastern United States seem low-risk. Things were different in Colorado, and judging by the look on his face after my response about a helmet, clearly my thinking was flawed.

A fellow software engineer, he proceeded to explain, "We gotta protect our heads. As knowledge workers, our brains are essential. It's the only way we produce work. I could figure out how to work without my hands, but I am useless without my mind. Trauma to our heads could jeopardize our entire career. No brain, no work. The potential fun is not worth the risk."

Our minds are our greatest asset. If you're reading this book, that includes you. Coming from outside of tech, this line of thinking was new. The jobs to which I was most accustomed involved manual labor. Previously, I worked in lawn care, dinner theater, physical training, and foodservice. They operated under different norms. Within physical jobs, hard work looks like long hours, sweat, and physical fatigue. Your hands are more important than your head.

In knowledge work, effort looks different. There may be no signs of outward exertion. There is little movement. The value of our work comes from the problems we solve. Our minds are what matter.

A year into my tech career, my dad visited our office. Hearing about the creativity and problem solving required in my work, he expected glitz and glamour. The reality was very different than he imagined. His interpretation was, "I thought I'd see people on whiteboards or immersed in conversation. Most of what I saw was just people staring at screens." Sorry, Dad, the work is going on in our heads. In technology, the outward appearance fails to represent what's happening within the minds and imaginations of those in front of the screen. Whiteboards and discussion have their place, but more often than not, effort looks uneventful.

Annie Liew, a graduate from an immersive coding program, drew even deeper parallels from snowboarding and knowledge work, "Like snowboarding, programming didn't come naturally to me. I remember all the falls, bruises and aches I had when I was learning how to carve.

"Your brain functions like a muscle too. It's important to treat knowledge work similar to the skill acquisition of musicians, artists, athletes and those who traditionally are known to train hard. If you're using it, stretching it, and growing, it will hurt. Congratulations, that is progress!

"As I exercise physically, my muscles may ache before they rebuild and become stronger. Likewise, the cognitive strain of trying to understand difficult concepts can feel challenging. You will fall and make mistakes. Get back up and trust the process."

Trust the process. That's difficult to do by oneself. Outside perceptions can make it even harder.

The underlying concept of knowledge work is *flow*[1]. Popularized by Mihaly Csikszentmihalyi in his book by the same name, *flow* is a period of deep focus. Our minds become consumed by the task at hand. It's a form of being "in the zone" and emerges as we think deeply about a problem or task for uninterrupted chunks of time.

Knowledge work, and the value it produces, are part of what contributes to higher paying software engineer roles. Writing code is valuable, but the broader expectation involves solving problems and overcoming obstacles. Software is one of the tools that enhances our ability to solve problems.

Consider the expression, "Work smarter, not harder." The phrase highlights the value of applying our minds, not physical labor, to a given task. If protecting your mind is important, it's also just the beginning. There are deeper facets of knowledge work. Some of those are foundational while others permeate each step of the process. A compliment to knowledge work is the value of continual learning. When our minds are our primary tool, learning is how we sharpen them. Learning isn't a task we complete. It is a ritual we must undertake regularly.

Speak with any expert. Even a few minutes of discussion will reveal the value they place on learning. Professionals are full of recommendations, references, and insights they have gleaned from the study of their craft. As knowledge workers, our value corresponds to the consistency and speed with which we pursue, acquire, and implement new information. Ignore learning, become obsolete. Embrace it and embark on a thriving career.

My trip to Colorado left a lasting impression. Long after the physical soreness wore off, the learning persisted. Lasting impact punctuates the nature of knowledge because it endures long after the moment that inspired it. Whether or not we realize it, our minds constantly store and catalog our education for later use. Like a snowball rolling downhill, when we apply ourselves to expanding our knowledge, our value and experience compound in size and speed.

HABITS AND RITUALS
Expertise in technology is not linear.

My (Morgan) first major in college was Business Administration. It was my first, but definitely not my last. Soon after, I changed my major to Exercise Science before switching again to Early Childhood Education. I went through college with zero emphasis on computers or software, dabbling in some of the furthest careers from technology. Who was I to think a transition into technology would be possible?

In 2011, around the time I left college, I started writing code. I felt behind immediately. The Dotcom Bubble had come and gone. Tech giants like Microsoft, Apple, Facebook, and Amazon were already established leaders in the market. The founders, engineers, and product teams in those companies, and countless like them, had decades of experience. I had none. The more curious I became about tech, the more daunting and unclear a career seemed. People were producing incredible, sophisticated work while I was barely making sense of the fundamentals. They were masters—I wouldn't even have considered myself an amateur.

If technology was a gold rush, everyone else was settled out West with mountains of gold while I was blindly wandering east. It felt like there were thousands of untamed miles of countryside between me and a meaningful career in technology. The phrase "I am so far behind," occupied my thoughts and fueled many fears with no sign of letting up.

It was then that I began considering exactly how far behind I might be. I started approximating. In Malcolm Gladwell's book *Outliers*, he mentions 10,000 hours are needed to develop expertise.[1] Plenty of writings have since challenged this belief, but I latched onto the simplicity of the idea: 10,000 hours of focus and deliberate practice lead to expertise. Without much thought, I sketched out some math. It would take three hours a day for almost ten years to reach 10,000 hours. Using a similar calculation, I could cut the time in half with six hours per day. With twelve hours daily, I'd knock it out in a few years. While daunting, the path to mastery started to seem more attainable.

My formula lacked nuance, for sure, but it also provided a helpful baseline from which to start. Even without a full-time job in technology, there was a path to expertise. Some paths were faster than others, but it was possible. If I could somehow align paid work hours toward the goal, success was inevitable.

I became fanatical about investing time writing code. I spent time every day. Whether early mornings or late nights, I would squeeze in time wherever possible. At times, despite tremendous effort, it felt like watching grass grow. To stay motivated, I looked for signals to track progress. You may be familiar with Github, a developer collaboration tool. Github maps daily developer activity in a graph with green squares. The darker the squares, the more attempts within a given day. I worked to create long, unbroken chains of dark green boxes. As months passed, I began waking up earlier and earlier to accumulate more hours before work. In the evenings, I got back online to do it again. The ritual of writing code was well underway.

Today, I'm quick to summarize those years of effort into a few paragraphs. If I'm not careful, my consciousness conflates ease with speed. It was a hard, slow process. There were thousands of hours, tens of thousands of error messages, and millions of lines of code. Early on, more moments were spent frustrated and confused than actually producing valuable work. For every line of code I produced, five lines were written and discarded. There was documentation everywhere but so much was outdated or presumed some basic understanding I clearly lacked. Helpful resources were scattered and disorganized. The pace was gruelingly slow and haphazard.

One day was particularly discouraging. I was reading through documentation for one of Google's many APIs. There were all sorts of abbreviations and jargon I'd never seen. After hours of pouring over highly technical specs I remember thinking, "Have I learned anything?!" Nothing I read seemed to make sense.

As months turned into years, I made a discovery. Expertise in technology is not linear. New technology regularly makes old learnings obsolete. Better tools and resources allow open-minded developers to skip steps. As technology advances, stronger skills can be gained with less effort. New updates and versions are released making previous headaches evaporate. Thousands of lines of code can be replaced with a single plugin.

The time I spent writing code was valuable, but the greatest skill I developed was that of consistent learning. Learning, combined with regular application, left a lasting impression. Digesting new information and deliberate practice are as much about the ritual itself as it is about the information.

Originally, I believed newcomers, like myself, were stuck playing catch up. At times, it seemed like it might last forever. But today, I realize the reality is more complex. While more seasoned professionals have amassed a wealth of knowledge and context, there are areas where newcomers have an advantage. Along with a fresh perspective, students entering the tech industry benefit from a consolidated understanding of the industry's twists and turns. Newbies seem more comfortable leveraging new tools and resources previously unavailable to earlier professionals.

There is a software term called "technical debt." When writing code, the choices and concessions an engineer makes along the way leave a trail of debris. Unlike financial debt, technical debt is unavoidable. Either due to changing priorities or technological improvements, technical debt can be limited but never fully avoided. Seasoned engineers carry a similar form of technical debt. I'll refer to it as experience debt. Sure, they have more years on their resume but many of the lessons they learned may become less useful. Experience debt accumulates by continually pushing work forward without spending ample time reevaluating old practices or beliefs. It's real and quite common among even the most experienced professionals. The digital world is rapidly evolving. If a software engineer isn't shedding old understanding and acquiring new skills, they become less relevant too.

Learning is about both the journey and the destination.

OPTIMIZING YOUR PERSONAL VALUE

I've worked hard over the years to increase my skills as a software engineer. I've also tried to add value to the broader engineering community. Some things have worked well and some have been utter wastes of time. Getting started, it's hard to know which is which. Lacking a traditional education in computer science, the following habits have allowed my career to grow and evolve more consistently than anything else. I've also had the privilege of watching these habits do the same for others.

Ask great questions.

Failing to teach active listening is the greatest oversight of many academic institutions. In my experience, neither grade school nor college offered courses on asking better questions or being a better listener. It is a priceless skill. The more I learn and grow, the more valuable I've found the experiences of others. Few things will accelerate the growth of a software engineer more than connecting with other talented people and leaning on their insights. Learning from your own failures and setbacks is valuable but the path to wisdom comes from avoiding hardship by learning from the mistakes of others.

These are helpful techniques as you identify and meet with mentors, coaches, or teachers:

- Prepare questions ahead of time.
- Take thoughtful notes as they share.
- Take action on what they say.
- Follow up afterward with gratitude and outcomes based on their feedback.

Experiment constantly.

Some days, I'd spin up two or three new projects to try out a certain feature or approach. It was tedious at times but caused me to get more efficient and deliberate. Years later, I have hundreds of projects from which to pull information and learnings. I rarely encounter a problem for the first time because of those early years of experimentation.

A few examples of ways you can experiment:

- *Find a website or app online and recreate it.* The popular social network of the day can provide a good starting point.
- *Research features or extensions within a framework and try to replicate them.* Building readable urls is an exercise I remember enjoying.
- *Find tutorial sites from trusted sources and go through each step.* Railscasts.com by Ryan Bakes is a personal favorite.[2]

I worked to further my mindset when eventually moving on to paid projects too. I pushed myself to say yes to many projects and challenges I didn't have all the answers to. I would give discounts and pad timelines by 200 percent to have the margin to stretch myself and my experience. This provided a safe environment to learn, while still allowing for real world experience.

Share your knowledge.
More often than not, teaching others is the best way to solidify your own understanding. There is no shortage of junior programmers seeking help. Regardless of how little you may know, try teaching someone else. You could try teaching a niece or nephew an introductory program like Scratch or volunteer with a code school class you're a few steps ahead of. Use what you know to help others.

Building a habit of sharing your knowledge keeps you sharp and reinforces the attitude that made the internet great to begin with. Wikipedia, an online encyclopedia, is one of the largest sources of information and is maintained exclusively by volunteers. Most of today's prevailing frameworks are also open source tools supported by unpaid enthusiasts.

If all else fails, write about your experience. Whether or not anyone reads your writing, the practice of organizing your thoughts leads to a more organized mind. To simplify the entry into writing, I set up a Medium account. It took seconds and allowed me to begin writing without worrying about style and aesthetics. Of course, I could have built a blog site from scratch. Actually, I had started the process many times before, but that became a distraction from writing. It was a form of procrastination. When writing is the goal, using an existing tool allows you to get right to the point.

Invest time daily.

As soon as you stop learning, your skills decline. In a fast-changing industry like technology, there is little room for becoming complacent. Early on, I found space whenever I could to read, watch tutorials, and interact with other engineers. Developing the discipline early pays dividends over time. The more I've challenged myself to be a lifetime learner, the more opportunities have followed.

Whether you're connecting with others, experimenting, or sharpening your skills, consistency is key. Tactical ways to make sure you're moving forward include:

- Scheduling time each day to explore new challenges.
- Scheduling time each week to reflect on what you've learned.
- Scheduling time each month to organize what you've accomplished.
- Challenging yourself to publish learnings (this will further sharpen your insight).

As a bonus note, waking up early is the easiest way I have found time in my schedule. Mornings provide fewer distractions and less mental fatigue.

If you're looking for a silver bullet to your dream future, I have nothing for you. While your mileage will vary, the road ahead requires consistent commitment and dedication. Much like writing great software, it takes time and it's rarely done alone. But action leads to opportunities. I've come to refer to this as "manufactured serendipity." Others call it luck. Either way, I've noticed a pattern among those who experience the most of it: they are active and looking for opportunities. The habits and rituals built into your daily life will feel like incremental moves, but the overwhelming impact of them can be just that: very good luck. What can you tweak in your daily routines right now to optimize for success? It's the smaller battles that win the war, and this is the starting place for aligning your life towards what's coming next.

YOUR UNIQUE "WHY"
That's right, I didn't get paid with money.

Why are you embarking on this career path? Take time to contemplate that question. As you do, imagine your ideal job. Get specific. Here are a few prompts to help:

- How much money do I want to make per year?
- How many weeks of vacation do I want per year?
- How many days do I want to work per week?
- How many hours do I want to work per day?
- How many more years do I want to work?
- How much health insurance coverage do I need?
- How much do I need to contribute to retirement?
- Do I want to work onsite or remote?
- Do I want to manage people or be an individual contributor?
- Which requirements are must-have?
- Which requirements are nice-to-have?

Everyone answers these questions differently. Every job provides different answers as well. These answers will change over time, and your first job will likely have different expectations than your next job. Your definition of success will impact the companies you pursue and the offers you can accept.

For code school graduates, I encourage a "learn before earn" mentality. Learning before earning means prioritizing opportunities to grow and stretch our knowledge rather than maximizing salary or benefits. Jobs with early-stage startups or small businesses, for example, may pay less but are generous with responsibility and more forgiving of mistakes.

My first real opportunity in tech came from a coffee company. The compensation? Free coffee for life. That's right, I didn't get paid with money. Despite not drinking coffee at the time, it was a great deal. The painfully low compensation afforded me lots of opportunities to try new things, make mistakes, and learn. As I proved my value over time, my responsibilities and pay increased. I was able to leverage the experience from that season of life into the next, larger opportunity.

Working for free isn't an option for everyone. Starting in my twenties, unmarried, without kids, and crashing on a friend's couch simplified my monthly expenses. Few have such flexible lifestyles. But anyone starting a new career needs to be ready to compromise in certain areas, or risk waiting and searching longer for the ideal opportunity. My standard for the right opportunity was low, so I seemed to have plenty of options.

Geraldine Galue, a code school graduate and friend, knew her why earlier than most. "I was passionate about nonprofit work. I knew nothing about coding, so I joined a bootcamp and was able to transition into a career as a software engineer at New Story, using technology to house families in need.

"For me, I genuinely care about working for a company driven by mission and values. There are companies where you can make more money or gain more notoriety, but it's less interesting to me. Yes, money is great but when applying to several different roles, I always sought companies involved in purposeful work. I look for work that makes me happy and where real-world problems are being solved. When discouraged, having a clear why gives me the resolve to push through."

A former mentor would regularly state a simple equation, "expectation minus communication equals frustration." Whether you realize it, you have expectations and assumptions. You may lack the flexibility of my early years or Geraldine's clarity, but understanding specifically what you want and need from a job will help you make more informed decisions.

Code School

IMPOSTER SYNDROME
Cool stickers on a laptop or water bottle don't translate to knowledge.

Imposter syndrome is one of the first mental roadblocks you will encounter entering your tech career. The goal of this chapter is to help you identify imposter syndrome, process the internal tension, and move forward.

What is imposter syndrome? Harvard Business School defines imposter syndrome as, "feelings of inadequacy without evident success."[1] I don't like this definition, mainly because evidence and success are highly subjective words. Who gets to determine if the evidence is sufficient? Who is setting the definition of success? I guarantee we would define both of those words differently than an Ivy League university.

I prefer a different definition. Imposter syndrome is an assumed gap between our skills and others' expectations. It's assumed because we arrive at our conclusion without actually asking anyone. It is also unlikely we even qualify to self-evaluate our skills and abilities. Humans constantly undervalue our own skills and overvalue the expectations of others. It's equivalent to comparing the weakest parts of ourselves against the best parts of others. The result is a distorted perspective.

Imposter syndrome appears, and reappears throughout code school. On your first day, it's normal to assess the others you will learn alongside. Few things are as obvious during in-person sessions. You'll notice someone who "looks like a programmer." Someone else will likely have techie stickers, like Github or Firefox, on their laptop. There is bound to be a group chatting in the corner about programming languages they use. Each of these things might cause you to feel inadequate or like you don't belong, but I'd like to begin with an observation: You all signed up for the same course. You are no more or less qualified than anyone else in the room. You each come from different backgrounds, but everyone arrived at the same destination. Some may have dabbled with code in the past and others are diving in for the first time. No matter where on your journey, you have made roughly the same decision. You all committed to learn and delve deeper into a tech career.

PERSPECTIVE SHIFT

The sooner you can shift your perspective, the better. You will bump against imposter syndrome throughout your career. It's unavoidable but it doesn't have to slow your progress. Lean into those feelings and open up dialogues with your peers and instructors. Jump into online forums. Ask questions. If you stay locked up in your own head, things won't get better.

You have more in common with other students than you have differences. Terrence Jackson shares a helpful perspective about imposter syndrome. "Don't track your progress against your peers. Some relative measurement is fine, but some folks just get it and others take a little extra time. In my experience the developers who struggle but stick with it will develop a host of other skills (debugging, research, human interactions) that propel them ahead of colleagues that seem to quickly get it."

Toby Ho, a skilled engineer and strong advocate for teaching software skills, recommends an alternative perspective for those he coaches, "I refer you to the work of Carol Dweck. She has a helpful concept called the *growth mindset*.[2] Her book, *Mindset*, is influential in all of psychology and it was life-changing for me. Her TED talk is a great intro to the topic. My one-sentence summary of how to combat imposter syndrome would be to accept that you are not there yet, but believe you can get there over time."

Remember, when you feel intimidated, you're not the only one experiencing these thoughts. To stay oriented, here are a few things to keep in mind to keep your head above water.

Ask, don't assume.

When you hear someone referencing past experience, become curious. Don't assume they know more than you or that they are further ahead. Treat the interaction like a learning moment. They might have a different learning style than you do. They might have to work much harder than you to learn the same thing, or vice versa. Instead of assuming, ask them about their experience. Ask them what they learned, how they learned, and where they learned. Lean into how you can use their advice to advance your learning.

Stickers mean almost nothing.

Cool stickers on a laptop or water bottle don't translate to knowledge. Stickers are cheap. In fact, you usually get stickers for free by attending any meetup. Type "sticker mule sample pack" into Google, and you will find a load of the most popular stickers for $1. They are a signal of your interests, not your ability as a software engineer.

Stickers are a good way to express interest in tech—that's all. If you see someone flexing, ask them about it. You might hear an interesting story or learn about a new meetup in town. If a sticker is unfamiliar, you might discover a new online resource or neat plugin.

You're struggling and so are they.
As an instructor, I would have students come up to me and ask if they were the only one struggling with a concept. That was never the case. In fact, a majority of the class struggled through each concept. Some hid their struggles better than others. The problem is no one benefits when classmates hide their struggles.

For example, consider the topic of "inheritance" in JavaScript. When you are struggling, which you likely will, the rest of your class is probably feeling it too. You will likely be at different points of understanding, but it's no less challenging. You may be wrestling with "object creation" while others are on "prototype chain." Heck, there is a good chance neither of these terms make sense right now. That's normal.

Either way, you're all struggling through the same concept. Rarely does any one person grasp a concept in its entirety. That's why it's so important to work together. If you understand one part and someone understands another, spend time together. Work together on a problem and craft a solution.

In traditional forms of education, collaboration is rarely encouraged. Group projects are the exception, not the rule. In programming, excluding blind copy/pasting, working together is a powerful tool.

Imposter syndrome is real. It can appear at different stages of your tech career. It's not a question of if you will experience imposter syndrome. Instead, invest in learning the techniques above, or creating your own, to work through it. If you find it immobilizing and the strategies you know aren't helping, ask for help. Be humble and speak up.

Lastly, remember what imposter syndrome feels like. One day, you'll find yourself at the top of your game and may notice others who are struggling. Maintain a sense of empathy, knowing you too were once in their shoes.

6

CODE SCHOOL CONSIDERATIONS
All resources are limited, including time.

Congratulations, you have decided to pursue a career in technology. It's time to learn the skills. Like most new skills, there are 101 ways to go about learning. But the most effective path to success is through an immersive program. Whether you are learning a spoken language or a programming language, similar principles apply. The easiest way to learn a different language is to spend time in a culture that speaks it. Immersing yourself will allow you to learn faster than picking up a few words here and there, and will keep you using what you've learned. This is one of the main reasons why code schools have popped up all across the world. Code schools provide a safe, learning-rich space to soak up the world of programming.

This chapter will not help you pick the perfect code school, because one doesn't exist. Discard the idea there is a flawless choice and relieve yourself of the pressure to discover it. Instead, learn about each factor to consider and which tradeoffs are most valuable to you uniquely. This chapter will guide you through teaching styles and options and with a little luck, you will find a few choices that fit your current lifestyle.

LIFESTYLE CONSIDERATIONS

When my career began, I was single, without kids, and sleeping on a friend's couch. Today, I'm married with children. The simplicity of the previous sentence fails to capture how complex my life has become, but family and other life obligations seriously impact your other commitments.

Before we jump into the types of code schools, let's first talk about you. Who are you? What works for you? No matter the choice, code school will affect your lifestyle. You get to decide when, where, and how much.

When I (Tim) taught at an immersive school, we made sure to over communicate the real cost of attendance. Beyond finances, time and effort rank among the largest costs of code school to a student. We told students to forsake their friends and recreational activities during the program's twelve weeks. Given the workload, there wasn't time for anything else. Those who listened, did well. Those who thought they could have it all, did not. In some cases, over-committing caused students to not even finish the course at all. This might sound extreme, but for most folks, this is the best method. Most programs are intense. You are signing up to learn a complex subject in a small amount of time. There are simpler, slower ways to kick-start a career, but they come with their own costs. Failing to fully commit can lead to disaster, but the benefit can be astounding.

Friend and director of multiple code schools, Toni Warren, says the following about the opportunity that comes from attending code school: "There is such a wide variation of folks that are successful. From never using an Apple product to then self-learning for years. How they perform is often about their support system, time they have to invest in themselves, and mindset they bring with them. But most importantly, they may want to try it before they actually decide to pursue it as a career. But hey, in what other industry can you do that?" The ability to "try before you buy" is another reason the tech path is valuable.

Everyone's lifestyle impacts their decisions. I've witnessed it firsthand in my own life. So, what's the first question you should ask? Consider how much time you have to invest. Right now. Do you currently have a job? Is it full-time or part-time? Are you married? Do you have kids? How many people depend on your time, attention, or income? How much are you willing to sacrifice to ensure success? The answers to these questions matter and will provide clarity as you approach future decisions.

TYPES OF CODE SCHOOLS

My first bite of Peachy Pistachio Chobani yogurt with dark chocolate pieces shattered my taste buds. Never before had I seen such a wild combination mixed into something so delicious. I've met friends who agree, while others hate it. Code schools are similar. They come in a variety of flavors, each with varying popularity and effectiveness. There is no one-size-fits-all solution. What's right for you will not be the same for everyone else.

There are too many variations to cover in this book. Instead, we will nail down a few staples. As you read, work to reflect on your previous education experiences and let them guide you. When did you learn best? What teaching styles were present? When did you struggle? Get out your sketchpad and let's find your path.

Consider full-time immersive.
This is the most common experience offered by formal code schools.

The effectiveness of an immersive code school model is directly correlated to an all-encompassing nature. They boast the highest success rates and facilitate an environment where you eat, sleep and breathe code. Your peers walk side by side with you in what's typically a Monday through Friday schedule. And the curriculum is speckled with a myriad of lectures, workshops, guest speakers, and collaborative projects. These types of schools range from 12 to 20 weeks—and you probably guessed it, the price tag is among the most expensive. But you get what you pay for. If you're committed and interested in the most focused route to your destination, this is the high-speed rail.

It's not easy to juggle an immersive program and maintain life's extracurricular activities, but many count the investment as worth it. Students who "clear their plate" before joining do much better than those who don't. I once met a student with five children who moved out of state to attend school. He and his wife agreed they would sacrifice in the short term in exchange for a better future for their family. Regardless of life's obligations or demands, it comes down to planning. Students who are intentional about preparation before starting perform better.

Consider part-time immersive.

Though most early programs were immersive, learning to code is not one-size-fits-all. Over time, schools pivoted and launched part-time immersive programs. Still in person, part-time courses stretch similar content over longer periods of time. Contrasting the 12 to 20 weeks of full-time immersion, part-time students may spend six to eight months in class.

One advantage of part-time programs is the variety of schedule offerings. Some programs occur on weeknights only. Others may be weekly plus Saturdays, while some are only on weekends. The goal is to provide a more natural fit to people's often hectic schedules.

If you still need to maintain a full-time job while learning to code, a part-time program might be the best fit for you. Some employers even have a continued education budget for their employees. Like any decision, it's not without drawbacks. Learning is hard when there are gaps between classroom days. All resources are limited, including time. And unless you're very disciplined, these gaps lead to days where you don't touch code. Think back to the foreign language analogy. Imagine bouncing in and out of another country throughout the week. Every delay and transition has a switching cost. You can still learn, but it will take longer to pick up the nuances of the language. It's possible, but will require you to take advantage of in between time to study and maintain your progress.

Let's pretend your course is Monday, Tuesday, and Thursday. Finding time to practice and write code on Wednesday and Friday will be essential. Nights, mornings, lunch breaks... any time you can manage. The success of your study will depend on your ability to do so.

Consider online.

Almost all human knowledge and understanding exists somewhere on the internet. You can look up anything you want to know. Why then would people pay for school online? Curation and community. When everything is available, the value exists in knowing what, how, and with whom to learn. Online programs provide the curation and community engagement that complement the learning experience.

For some, attending a school in-person is off the table. Whether it's because of location, lifestyle, or some other factor, it might be impossible. Unless you live in a major city, it's unlikely there are in-person programs in your area. In this case, a lot of code schools offer online programs. These options provide a mix of learning sessions. Session types might be live, recorded, or a mixture of the two. There are presentation sessions and opportunities to live program alongside a group.

If you choose an online program, benefits reflect a part-time immersive program. But consider the many differences too. Learning while watching a video is oftentimes harder than engaging in person. With online programs, it can be difficult to ask questions in real time. In-person experiences curate an environment to help students focus and interact. When online, however, that is your responsibility.

What an online program gives in flexibility, it costs in the demand to structure and prioritize your learning time, space, and behaviors. Learning is difficult enough without distractions. Learning to code with distractions may be impossible. If you decide to do an online program, planning goes a long way toward maximizing the experience. For recorded sessions, I suggest writing down questions to ask later as a helpful technique to keep up during online sessions. When remote, create a physical space where you can focus during programming sessions. And work to minimize the number of days without practice. Each improvement compounds, but only when you invest consistently. Consistency and focus increase your odds of a positive outcome.

Consider self-paced online.
The last learning style we will cover is self-paced learning. Self-paced programs, until recently, were generally not even offered by code schools. As of 2020, some schools began to offer these as a way to give students more options. Before schools offered them, online courses were the main alternative.

Instructors, teaching assistants, and peer cohorts differ in self-paced programs. Depending on the code school, these compliments are very fluid or nonexistent. While not immediately obvious, it's helpful to work through complex concepts with others. To account for decreased connection, schools offer online forums and chat. Some even provide ways to hire a mentor. It's helpful but these tools struggle as a true replacement. I suggest utilizing them as a tool to sample the playground rather than learn job-ready skills.

Self-paced programs may be your stepping stone. While they fail to match the speed or support of their counterparts, they can help ensure software is a fit. Most people start excited, but interest may wane so self-paced programs are excellent for testing the water. You can sign up for a lower cost, try different programs, and make sure it's right for you.

Persistence, thoughtfulness, and discipline are essential, regardless of the route you take. Fast, easy options do not exist. Whether you are doing immersive, part-time, or self-paced, you need to make a plan. You need to decide how you will tackle your days, nights and weekends to maximize your learning. Limiting and removing distractions is especially helpful. Try to write code as many days as you can and pay special attention to consecutive days.

Spoiler: Code school is the beginning of rigorous learning, not the end. When you finish your program, getting a job and building a career require similar intensity.

PRICING OPTIONS

For the first code schools, pricing was simple: a flat fee for a specific period of time. Pricing worked a lot like traditional college institutions where tuition is paid up front. Over time, many school creators realized this was a hard business. Difficulties compounded because few code schools qualify for federal loans. Third-party lending became popular, but it resulted in a general sense that school was being paid via a credit card.

Today, a lot of those early kinks are resolved. Much like the type of school you attend, financing is an important part of the decision-making process, and students today have options. The following financial options are among the most common.

Explore flat-fee.

It's worth mentioning that many schools still provide and promote flat-fee pricing. A flat-fee is when you pay a specific amount for your multi-week program. Most options are cheaper than traditional colleges. They also provide the prospect of getting a job much quicker. Many folks still find this a viable option.

One mistake I often see students make is underestimating the true cost. You will need money for tuition, living expenses, books, programs, and other resources to fuel learning. I equate it to college room, board, books, and discretionary spending. These items are important, but not part of tuition listed on the school website.

Explore loans.

No matter the type of school you attend, there are companies willing to loan money for costs. Given a rise in popularity, companies now exist specifically for coding programs.

A powerful feature of many such loans is interest-only payments. Because these programs provide future employment, lenders allow you to defer principal payments. This defers your costs in a major way. Some even provide interest-only payments until you land a job or for a fixed amount of time.

These are helpful terms when considering a loan. Debt can impact your short-term and long-term life plans, in positive and negative ways. Keep in mind, the loan will need to be repaid whether you finish the program or not. Students find themselves burdened with repaying a lender while trying to look for a job. Many end up working back in their previous field, but now with loan payments.

Explore income share agreements.

Income share agreements (ISAs), have become a very popular way to pay for a code school. The concept is simple. You attend the program for free and only pay when you graduate and get a job. Once you have a job, you pay back an agreed upon percentage of your monthly paycheck until you hit the program's cap.

This payment method is very attractive. ISA's allow you to attend school without an upfront cost. It has the added benefit of putting more pressure on schools to assist in a student's hunt for a job. When a school gets paid upon job placement, their incentives are more in line with yours. If a student doesn't find a job, the school never gets paid back.

In traditional educational institutions, less than five percent of costs involve post-graduation support. If you attended college, think back to that dingy career-services department. Pushed aside in university basements, they fail to provide value in proportion to the steep tuition costs.

Income share agreements are not perfect. It's helpful to defer payment but the total cost could be double that of paying upfront. You should also consider any pay gaps between your current situation or career and your entry level tech job. Many students transition out of one career to pursue engineering. Plenty of people leave high paying jobs in pursuit of more happiness or fulfillment, but if a percentage of your entry level salary must be repaid (sometimes up to 20 percent) to your school, living expenses could be a challenge. This arrangement can also limit job opportunities by requiring a minimum salary. In order to ensure pay back, ISA conditions are built in to repay.

Even so, ISAs have become the go-to for many programs and students. The terms are changing regularly and seem to be more in favor of students. If you decide this is for you, make sure you check out all your options and understand the fine details of the agreement before entering into it.

Explore scholarship opportunities and the GI Bill.
When the school I (Tim) worked for first launched, we always gave two students in each course a full scholarship. We did this to both fill the classroom and to offer scholarships to those who couldn't otherwise afford it. We never advertised these scholarships and only offered them to students who were declined for a loan or didn't seem like they could make it work. And that leads us to the vaguest of financing options: scholarships and stipends. While many code schools traditionally don't offer scholarships, some are starting to see the benefit. Whether it's a diversity scholarship, one that's offered for a new program, or something specific to that program, you'll want to make sure you look into all of your options. The big point? Ask. Especially if you don't see any opportunities advertised.

Lastly, a lot of schools have been working hard to accept the GI Bill, a government education program for military veterans. If they do, they will usually advertise this, but if not, it's always something to ask about if you qualify for it. In many cases, a code school is considered in the same category as a trade school, which will likely make it qualify, but it will be a requirement of the school to apply and complete all the necessary steps to accept it.

Like my favorite Peachy Pistachio yogurt, code schools come in many shapes and sizes but the differences don't come without tradeoffs. I encourage you to pause here to mark out your own considerations and capacities thinking of time, finances, and current obligations. Once you have a clear picture of you, the following chapters will prove doubly valuable.

7

MAXIMIZING YOUR CODE SCHOOL EXPERIENCE
If you wait until everything makes sense, you'll never get started.

I started my career before code schools became popular. In hindsight, I struggled more than was necessary. The opportunity to accelerate learning and become immersed in an intensive experience can shave months or years off the learning process. I benefited from helpful friends, but so much of my time was spent parsing through outdated techniques and fumbling around the internet. Code schools have emerged alongside the natural maturing of the internet. There have never been more opportunities to learn than now.

YOUR EXPECTATIONS

It's about more than just the school experience. Riaz Virani emphasizes the importance of preparation before starting code school. Today, Riaz is a Chief Technology Officer at a startup and organizer of Atlanta's local Ruby meetup. He's also a code school graduate. Riaz advises, "Focus more on what you should do before starting a coding bootcamp. I'd recommend doing a boot camp after you've already attempted to learn via Udemy, Udacity,

FreeCodeCamp, or other online resources. It's tough to start from scratch and get up to a level where you can be productive within a few months. We've hired two boot camp graduates recently at our startup and both spent three to six months of their free time learning before they started boot camp. It seemed to really help them succeed. The boot camps are great accelerants of learning but there are better ways to get your feet wet."

Whether you're preparing to start a course or your program is underway, the following behaviors will help you make the most of your experience.

Own your own experience.
Code school is not a magic bullet. The information doesn't always flow seamlessly from instructor to student. A tremendous amount of effort and mental flexibility is required to maximize the opportunity that code schools provide. Between your instructor, community manager, and peers, a wealth of knowledge exists but it's up to each student to make the most of it. The students who own their personal experience and take it seriously gain the most.

Learn how to learn.
Throughout school, I (Morgan) was a slightly above average student. I could be counted on to bring home mostly As, plenty of Bs, and every few years I'd spit out a C. What puzzled me was the inconsistency of my performance across subjects. There was hardly a trend among my best and worst grades. It was a mixed bag.

One thing I learned early on was that I had to attend class. If I attended class and paid attention, I could expect to perform well. Unfortunately, it wasn't until years after college that I connected the dots to auditory learning. When I hear information, I retain it five to ten times better than when I read it. Since then, I've become a voracious listener of audio content like podcasts and audiobooks.

There is no shortage of people who snub their noses at audiobooks, but they work for me. For me, it's not a question of ease as much as accessibility.

More important than *what* language or skills you learn is *how* you learn best—followed by the ability to draw connections between what you're trying to learn and what you already know. Learning how to learn is the most valuable contribution of any code school. They've done the work of making the knowledge practical and modern, but it's up to you to internalize the process. Considering how quickly technology evolves, adaptability is a skill in and of itself.

When confronted with unknowns, the strongest engineers I know have a process they use to acquire knowledge. It's different for each of them but they are comfortable learning new information. They invested time into understanding how they learn best and eagerly pursue knowledge.

Focus on good enough.
You're not going to achieve mastery in a few weeks. If you're hoping to nail everything at 100 percent, you're optimizing for the wrong outcome. With a bootcamp style experience, information will flow faster than you're able to retain. Drinking water from a firehose is a great analogy of the road ahead. Yes, you'll miss things and feel overwhelmed—that's natural. Gather as much knowledge and skill as possible. Don't sweat areas that go over your head or aren't crystal clear. Deliberate practice and experience have a way of bringing things into focus over time.

HTML was the first area I (Morgan) felt tension. I wanted to know what "div" stood for and why "class" meant what it did. I was flooded with so much terminology that did not make sense. This became a barrier to forward movement. Keeping it simple, I tried to focus on what information was required to complete the work at hand. With each new experience, I was able to add understanding to my comprehension. Much like breaking in a new pair of shoes, what once felt clunky and constraining eventually fit like a glove.

This seems counterintuitive. You are here to learn after all, right? Despite popular opinion, more information does not lead to better decisions. In most cases, our insistence on prioritizing information-gathering over action-taking is a form of procrastination and fear. It takes courage to act without understanding the full picture. As you learn, this tension is an important part of the process. Action leads to deeper levels of understanding, not the reverse. If you wait until everything makes sense, you'll never get started.

KEEP IT PRACTICAL

Making sure to provide some practical tips, here are a few pointers that should help maximize your experience:

Keep your folder structure organized.
Use thoughtful names and clean things up when they get overwhelming. Rather than deleting old code, create an "archive" folder. It's helpful when you feel discouraged to look back on past projects to see how far you've come. There is a practical reason as well. Storing your old projects creates a private library to pull code snippets of features you've built before.

I set up my computer with a "code" directory. Inside, I have a personal folder and one for each company I work with. Inside each of those, folders are labeled by project. The consistency makes it easy to store large numbers of projects, orient myself quickly to active projects, and navigate to previous work.

Schedule a time each day to practice.
This time should be in addition to classwork. Deliberate, independent practice is key to achieving mastery of anything. If you're expecting to learn everything within class, prepare to be disappointed. Whether in the evening or early mornings, set aside a time and place where you invest consistently in your own independent exploration.

Early on, I lived in a small, one-bedroom apartment. The kitchen table was my daily, early morning workspace. After waking up and going through my morning routine, I'd spend a few hours coding at the table before heading off to work. I believe this positioned my career early on toward continual learning and progress. I still go to my kitchen table in the early mornings to work.

Set up a network of skill levels.
Get to know a student that is more proficient than you, another who is at about the same level, and one that's a little further behind. The first (more proficient) student will be able to provide helpful insight you may be missing. When you get stuck, their assistance can help you overcome barriers. The peer will push you. The benefits of working through things shoulder to shoulder can create an edge over time.

The student who is step behind you will present an opportunity to teach. Teaching something you've learned, no matter how fresh, will refine your understanding of the material. Regardless of the experience level, focus not just on what they do, but how they think about the work. Remember, you're trying to learn how to learn more than just complete the assignments.

Keep a simple journal.
This framework is simple and shouldn't take more than a few minutes each day. It will keep your thoughts organized and can

provide great material to add to your portfolio or share with an employer. On paper, answer the following questions:

- What did I learn today?
- What do I hope to learn next?
- Where am I getting stuck?
- What am I going to do about it?

One of my favorite interview experiences began with the candidate sitting down and pulling out her laptop and notebook, with matching covers. As we asked her questions and assigned tasks, she went to her physical notebook first before proceeding to the keyboard. It was clear she wasn't putting on an act. Instead, she was constantly processing, documenting, and reevaluating her learnings in real time. Whether or not she was an excellent programmer then, thoughtful learners like that do not stay juniors for long.

8

LEARNING BY DOING
When you board an airplane, you have peace of mind knowing it's not the pilot's first flight.

I f you're in school, you've likely heard before, "Alright class, let's build a to-do list app." How many times have you built something that looks like some version of a to-do app? If you haven't heard this already, you will. Building simple, repetitive things like this can be frustrating, but it is one of the best ways to improve your coding skills.

Don't worry, this section is not going to be about the benefits of building a to-do app, but rather about the need to continually learn through taking action. People learn best by doing. Whether you want to pick up a new sport, learn a new musical instrument, speak a foreign language or get better at writing JavaScript, nothing compares to diving in and doing the work. There's a reason the apprenticeship model still exists, and why so many industries require internships or residencies. You need to spend time actively participating in a skill before you can become good at it.

When we go to the doctor, we assume it's not their first time seeing a patient or their first time administering a shot. When you board an airplane, you have peace of mind knowing it's not the pilot's first flight. Most software engineers aren't programming pacemakers or involved in high stakes industries, but the need for seasoned, battle-tested experience is no less important. Taking action will not guarantee expertise, but it is a requirement to getting there. It's important to spend as much time learning and doing the work as you can.

This feels like an appropriate place to insert a mindset shift from our friend Rose Lake, Software Engineer at Stitch Fix. Rose says, "Supplemental learning is paramount. Students can't expect their bootcamp instructor to give them all the knowledge they need. Expect to use YouTube, Stack Overflow, language specific docs, mentors, etc., to enhance your understanding of the material you're learning."

Experience, from every front, leads to expertise. Pop culture is quick to claim "natural talent" but the journey to greatness is often overlooked. Year after year, we hear stories of "the overnight success." It's a myth. Michael Jordan, arguably the greatest basketball player of his era, was the first to arrive at practice and the last to leave. The Wright Brothers, creators of the first airplane, spent years crashing into sandy dunes before their plane took flight. Thomas Edison conducted thousands of experiments before his lightbulb breakthrough. Getting good requires work. A lot of it. Anders Ericsson wrote an excellent book, *Peak*, on the topic of mastery.[1] Despite traditional wisdom, Ericsson argues that consistent, deliberate practice is the only path to expertise.

When I (Tim) first started learning to code, I used to walk to the library on my lunch break and check out books on programming. Back then, there were no code schools, no online learning platforms and not many places that would hire an unskilled programmer and give them a shot. I checked out books on HTML, Java, C#, Visual Basic, .NET and the like. I had no clue if these languages worked together. I simply knew that I wanted to be a programmer and these were the only books they had available. Over time, I would begin to understand front-end versus back-end and the nuances of the different languages, but in order for me to get there, I had to dive in. At the end of the books, which usually had challenges or practicing points, I worked hard to complete what I read. By doing this, I learned which languages I enjoyed working with, and which ones I did not. I began to understand how my brain works and how I learn. I then found websites like CSS Zen Garden where I could practice and continually improve my skills.

When I applied for my first job in tech, I had no clue that there were about 20 other folks trying to get that same position. I went in for the interview and they had us all lined up as if we were there waiting to read lines for a new hit TV series. One by one they called us back, asked us a few random programming questions and then made us fill out a multiple-choice form about CSS and HTML. It was terrifying. When my turn was done, I was asked to sit back down in the waiting room. About an hour later, they started calling a few people back at a time, and one by one they each left.

By the time it was my turn, my heart was pounding. They called me back and sat me down in front of the CTO, who offered me the job. I was shocked. He shared that the majority of applicants had done about the same on the quiz and had all interviewed well. The difference with me, he said, was that I was the only one who had a website. I was the only one who had proof that I had any experience at all in building something. Don't get me wrong, my website at that time was nothing special, but it was coded in HTML and CSS and did not have a single error in its code. When they realized their quiz didn't help them land on a specific candidate, they started reviewing everyone's resumes again. That's when they found I was the only one who filled out the website field, and they picked me. It's currently

2021, and if you tried to apply to a job without a website, it's likely you would not get a call back. What back then was a website, today is a Github portfolio of programming experience. When a hiring manager sees that, they know that even if you don't know the exact frameworks their team uses, you are at the very least someone who builds. You're someone who learns by doing.

9

ESTABLISHING YOUR NETWORK
Relationships are not about fulfilling your needs.

With some pvc pipe, glue, and hairspray, my friends and I learned to make a "toy" capable of launching an entire potato. We began with little more than a bored afternoon and some spare change. By the end of the evening we were sending spuds into the stratosphere. We aptly named our creations "Potato Cannons" and throughout middle and high school we made dozens. Years later, I cofounded my first business with one of those friends.

While trivial, my Potato Cannon days prove a valuable point: every major, positive life event is a result of my network. This theme remains consistent within my own life and others I know. For example, my first paying website build came from a colleague's introduction. The cofounder of my first business was a decade old friend. The largest deal I've closed was from a referral. My first opportunity as a chief technology officer came from a relationship with one of the cofounders. I even met my wife through a mutual connection. Looking back, I'm unsure of the quality or lift of my career without the impact and influence of a strong network.

YOUR GREATEST RESOURCE

A healthy network is a working professional's greatest resource. It's a combination of who you know and your reputation among those individuals. Beyond merely knowing large numbers of people, it's important to be well regarded by your connections. The simplest way to do this? Provide value. Consistently providing value is a proven way to build better relationships and strengthen your community ties.

A friend of ours emphasized the importance of building a network when we talked with her. "A professional network is a game changer," said senior web designer, Kelly Leonard. "If you go to code school and stick to yourself, you've lost half the value. So much of my work now is collaborating with others. It's imperative to the job hunt, and its value trumps even what you're learning."

My favorite analogy to describe networks is banking. While banks are not known for their personability, they possess overlapping elements with relationships: balances, deposits, withdrawals, and individual accounts. Wealthy people have lots of accounts and each account remains in good standing, with large balances. They make more deposits than withdrawals. Relational wealth works much the same way. A consistently growing, positive balance with more people leads to greater relational equity.

Much like saving money, organically growing your network can take years or decades. Fortunately, code schools provide students with a head start. Gaining access to an established network of skilled teachers, alumni, advisors, and employers is equivalent to a financial inheritance. It's an injection of relational equity you didn't have to earn yourself. You benefit from your association to your code school because it immediately gives you common ground for connection.

Leaning on an established network is a valuable tool to kick-starting a network of your own. This is commonly ignored. Every teacher, alumni, advisor, and employer have experience and insights to share. You can save yourself time and struggle by opting to learn from others. A grade school teacher shared my favorite quote on the topic of learning from others, which stuck with me. They said, "A smart person learns from his mistakes, but a wise person learns from the mistakes of others."

Here are four techniques I encourage young developers to commit to when engaging others:

- Provide value.
- Express gratitude.
- Stay curious.
- Ask.

Provide value.
Providing value is all about giving more than you expect to receive. Relationships are not about fulfilling your needs. Zig Ziglar, a famous leader and teacher believed, "You can have everything in life you want, if you will just help other people get what they want." This encourages a posture of helpfulness, listening, and generosity. Those known for generosity tend to prevail over time because this posture is rare and impactful.

Here are a few easy ways I look to provide value to others:

- *Ask how you can help.* One of the greatest closing questions to any meeting is, "What are one or two ways I can help you?" To improve the concept even more, take action quickly on what they say.

- *Make an intro.* Introducing someone you have met to another helpful person in your network is akin to professional gift-giving.

- *Summarize your takeaways.* In a follow-up email, summarize two to three points you took away from the conversation. It shows you were listening and reformats their insights into a format they can more easily pass along to others.

- *If meeting in person, get the check.* While it's unlikely to be a large sum, the gesture of instinctively paying for the other person has value. It's also a simple way to express gratitude for their time.

- *Send a gift.* Keep in mind, company swag or promotional material is generally not received as a gift. I've sent Tiff's Treats, Doordash, and other practical gifts as expressions of gratitude.

- *Give books.* Whether physical or digital copies, sharing books I found valuable and why I think it would connect with them makes an impression.

Express gratitude.
One month in 2016, I was particularly exhausted. Our business was struggling and it seemed like every project was off track. Swarmed with frustration and anxiety, I began writing LinkedIn recommendations for people I'd recently met. Nothing particularly long or insightful, just words of encouragement and gratitude. My attitude and outlook began to shift. The resulting thanks and responses compounded the effect. I'm not sure how many people read LinkedIn recommendations, but I can confidently say the recipient will. That's more than enough reason to do it.

No one owes you anything. The time, energy, and investment made by others into your life is a gift. People who operate from a position of gratitude garner more support and maintain a higher degree of satisfaction than those practiced in entitlement.

Stay curious.

Everyone has something to teach us. Unfortunately, insights are rarely packaged neatly. Traditional education puts people through thousands of hours of instruction often without teaching active listening, healthy discourse, or interview skills. Asking good questions and learning from the experiences of others can spark significant inspiration and understanding within our own lives, while also giving others the opportunity to share their learnings.

Ask for help.

It's unlikely people will start guessing what you need. Making an ask, while humbling, provides clarity to those within your network. Building upon the earlier "relational bank account" analogy, an ask can be likened to a loan. When done well, asking for help and honoring the generosity of others can actually build credibility over time.

There are a few types of "asks." An introduction is the most common form of an ask I've found helpful within my network. Introductions are a way to expand your network, broaden your understanding, and deepen your expertise.

Transitioning into tech, you probably don't have a pool of technologists from whom to learn. It's more likely however, you are one or two degrees removed from someone. Appealing directly to your current community can open doors to new connections within their network. My favorite approach is to lead with a question or a struggle I'm having. For example, I was trying to connect with educators in tech. I asked friends and colleagues, "Do you know anyone who teaches programming or mentors others?" After hearing their responses, I asked if they would make an intro. Within a matter of days, I met a handful of relevant, new professionals. The greatest part about introductions: they are reciprocated. When someone introduces you to a friend or a connection of theirs, ask if there is an introduction you can make for them. It works full circle.

While it can feel intimidating, engaging your network is actually you stepping into a moving ecosystem of professionals who all want to grow, expand, and meet others like them. It may take time to gain your footing and grow your confidence, but by providing value as soon as you can, expressing gratitude where possible and taking leaps where you identify them, you'll fold into the ecosystem before you know it. Things work better when people can see beyond themselves. Making this a foundational element of your career path will serve you well long term.

10

BUILDING YOUR BRAND
Action is what uncovers more opportunities. The only mistake is doing nothing.

One of the most widely known tech contributors in the last decade is Satoshi Nakamoto.[1] Not only did Nakamoto's ideas contribute to blockchain technology, but he is the founder of Bitcoin. So popular is his influence, the smallest unit of Bitcoin is named after him, a Satoshi. The best part is that no one knows who he is. That's right, the founder of Bitcoin is completely anonymous. Better yet, Satoshi Nakamoto is believed by many to actually be a group of contributors acting under a single alias. Their resume and work experience are irrelevant, because their body of work speaks for itself.

While we may never garner the same acclaim as Nakamoto, this story captures the beauty of technology. Our identities can be shaped and amplified beyond what's possible in the physical world. Our current employment and resume can extend our influence, but we are not limited by them. More than almost any other industry, the work of an engineer can speak for itself, or at least on our behalf.

YOU, ON THE INTERNET

Friend, code school founder, and business owner Richard Simms prefers to set realistic expectations. He councils, "You are not going to come out of a coding bootcamp and land a six-figure developer position at Google on day one. Honing your skills, building up a portfolio, and gaining experience is rarely enough. You need to build relationships, grow your network, and be ready to search hard for that first foot in the door. Getting your first job is going to be hard. As you build up credible experience, you will have more options and increase your earning potential. As you invest in yourself each day, your credibility grows and your nontraditional background becomes less and less relevant to folks that will consider you for their team."

The following suggestions when applied regularly, can help you develop a richer online presence. The first few suggestions are actionable steps. They're tasks that shouldn't take longer than an hour or so to complete. As we discuss contributing online deeper in the chapter, remember the importance of baby steps. Reputations are not built in a day. You do not need to know everything from the beginning. Circumstances will change, the industry evolves, and new opportunities seem to appear overnight.

Annie Liew, who you may remember from the chapter on Knowledge Work, shared a comforting insight. She compares getting started to driving in the dark. "I love the analogy of driving at night. You cannot see what's beyond your headlights but you trust that you're going in the right direction. As you keep moving, more of the road is revealed." Action is what uncovers more opportunities. The only mistake is doing nothing.

Buy your name as a domain.
It is helpful to own your name as a domain. It is a great place to park

your portfolio site, and Google places a higher value on domains with a close match to any search term. Why does this matter? A Google search of your name can be the best portfolio piece over time.

Try googling the authors of this book: "Morgan Lopes" and "Tim Whitacre." Assuming Google hasn't matched your search to a relative with a similar name, you are likely to see our personal websites within the top five results. It's not because we are prolific online celebrities, it's simply because time and consistency are on our side.

Grab your social profiles.
Credibility on the internet is built upon backlinks. Backlinks are sources that link to your name and likeness online. The more credible the source, the higher Google values the backlink. Simply put, more backlinks mean more visibility. The easiest way to start building backlinks is by claiming a consistent username across popular social channels. Whether or not you intend to engage sites like Facebook, Twitter, or TikTok, there is value in owning your username and likeness on those platforms. Not only will it deter others from "squatting" on your likeness, it adds more connections to areas you do choose to invest.

Here is a starter list of online profiles to consider grabbing. Keep in mind, there is value in maintaining consistency in terms of name, username, and profile photo:

- Facebook
- Instagram
- Twitter
- YouTube
- LinkedIn
- Github

- Stack Overflow
- Reddit
- Product Hunt
- Medium

Struggling to find a consistent username? Google "username checker" and you'll see a number of tools you can use to set different usernames across many sites. It's unlikely you'll find one that works everywhere, but it'll help you get close.

Mirror what others are doing.
When in doubt, investigate what others are doing. Instead of focusing on their habits today, look into what worked for them in the early days of their career. Our friend Kelly Leonard shared great thoughts around mirroring others. Kelly said, "A helpful hint I give people is to find someone who is doing the job you want. Check out their LinkedIn, portfolio website, and resume. See what it takes to be where they are and make a checklist for your journey. Maybe they talk about a certain program they use or things they're learning. Or maybe you find certain talents they highlight in their resume. It's important to maintain your unique edge. It can also provide inspiration to spruce up your resume."

Package your story.
Platforms, profiles, and portfolios are valuable, but an online presence that effectively communicates who you are and what you can do is invaluable. This doesn't require incredible skills, but it will require intentionality in how to package your story. I return to the following frameworks every few years to help refine and clarify my personal story. Pick one, try them all, or settle for something in between. More valuable than any one framework are the lines of thinking they provoke.

- *Core Values.* These are the underlying principles that safeguard your decisions. My core values, borrowed from the first company I started, are: create remarkable experiences, pursue excellence, not perfection, be effectively human, and go boldly forward.
- *Purpose Statement.* When presented with two great options, how will you choose between them? A purpose statement is a clarifying phrase that summarizes your life's purpose. My personal purpose statement is to "pioneer a more beautiful future for others."
- *Mantra.* Throughout history, armies carried a chant into battle. This unified them and steeled their resolve when things inevitably got tiring. My rally cry is, "Show them." Instead of expecting people to believe me, or trust me, I should be prepared to show them what's possible.
- *Venn Diagram.* Imagine the contexts in which you specialize, enjoy, or excel. Given the world's endless professions and choices, there is probably a profession or industry where all of those overlap.
- *Super Power.* What is your unique competitive advantage? If going head to head with competition, where do you shine? My super power is rapid prototyping and experimentation. My creativity, passion, and interest are unlocked trying to problem solve during the early stages of an idea, business, or project.
- *Lifelines.* A lifelines exercise involves mapping the high and low points throughout your entire life. Every significant moment you can remember. Afterward, reviewing and evaluating those key events can illuminate patterns, beliefs, and experiences which might be shaping your choices.
- *Annual Reflection.* Time is our most valuable resource. It's the only truly scarce resource. Every year, over 8,000 hours

come and go. Most adults move from year to year with little regard for how that time was spent. Since 2016, I've spent a few hours each year reflecting on the previous year and planning adjustments for the coming one. It's more than New Year's resolutions. It's an application of engineering principles within the life I want to live.

- *My Precious Moment.* Thinking about the future can be overwhelming. I've tried to simplify the discipline by imagining a single future moment instead. For almost a decade, it's been the same moment. I imagine as much desirable detail about that moment. The goal is to paint a clear picture, then work backwards to figure out what must be true for that moment to exist.

- *Ikigai.* This Japanese concept is a variation of the Venn Diagram mentioned above. Instead of choosing what the circles represent, Ikigai uses four overlapping circles: 1) What I love, 2) What I'm good at, 3) What I can get paid for, and 4) What the world needs. As you fill in each circle for yourself, it's believed your sweet spot exists where all those elements overlap.

To read a full guide on each of the above, visit codeschoolbook.com.

Each time I work to refine my own story, which is constantly evolving, I learn something new about myself. The clearer I'm able to articulate my story, the more appealing I become to certain groups of people. Today, my goal is to live at the intersection of entrepreneurship, technology, and social impact. Work opportunities that align closest with those contexts are uniquely exciting to me. Through articles I write, LinkedIn posts I share, and relationships I invest in, the clarity repels disinterested individuals and becomes a beacon, attracting others with aligned interests.

Everything does not have to revolve around tech. It's about exploring what makes you uniquely valuable. In most cases, there might not be an obvious tech connection at all. That is fine, as long as it's genuine and you are able to communicate why others might find that valuable.

Richard Simms put it like this, "Code school grads should also think about their past experience and skills and how to speak to those to paint a fuller picture of what they bring to the table. Oftentimes, particularly for code school graduates, a potential hire can look more promising if there are other ways for them to bring value aside from strictly developer skills." Tell your story. If you have past experiences in another field, put that experience to work for you.

Contribute online.
After you lock in the basics, there are many ways of further bolstering your online presence. The following list provides unique ways to expand your presence online. Everyone is different, so instead of trying to replicate what you see others doing, find something that fits your unique personality and style.

- *Write.* Whether you prefer a micro-blogging format like Twitter or something long-form, writing helps clarify thought and leave a train for others to follow. Yes, even early stage professionals have insight to offer. There is always someone more junior than you who might benefit. At the very least, your own mental clarity is enough of a reason to write.
- *Assist open source projects.* Open source is the backbone of the internet. Take a look at your favorite libraries online and chances are they have an open list of issues waiting to be tackled.
- *Sign up to speak.* Giving talks, much like writing, makes us more effective communicators of our ideas. There is no

shortage of online and in-person events to speak at. You don't have to desire a speaking career to reap benefits from speaking, you just have to have a valuable message to share.

Your online presence informs your overall trajectory. I've heard students express fears of embarrassment and being pigeonholed while discussing investments in their online presence. These fears and others like them become immobilizing and limit long-term potential. Careers struggle more from obscurity and ambiguity than over-exposure. Hiring managers, like most people, like what they can perceive. Over time, it is natural for your interests and online expression to evolve. Do not let the potential of future change prevent you from getting started today. A friend of mine once said to keep up the first blog posts you've ever written. It's helpful to see how far you've come. The ability to show that you've been chipping away for an extended amount of time is more valuable than perfection. Colleagues and employers know that.

So, start with what you have. Confidently place your name on the internet where you can. The consistency will compound and will take you further than you think.

Job Hunt

11

THE EXPERIENCE DILEMMA
Employers require experience before they'll hire you, but how are you supposed to gain experience?

Pretend you need a pair of scissors. You search online for the sharpest pair you can find. You read dozens of reviews and can see which scissors are clearly the best. Days after your purchase, a package arrives. As you open the box, you realize the scissors are protected in a dense, plastic case. It is sealed from all sides. You pull, twist, and pry but the case doesn't budge. It seems you need scissors to cut through the case. Unfortunately, they are locked inside.

Round and round the example goes. The case remains unopened because it requires the tool inside. You are stuck.

Starting a career is similar. Past work seems to be a prerequisite for new opportunities. Employers require experience before they'll hire you, but how are you supposed to gain experience? If everyone needs experience, where does anyone get started?

We call this the Experience Dilemma.

From the employer's perspective, every hire is risky. Regardless of how great the interview process, hiring is never a sure bet. So, employers use past experience to inform their decisions and reduce their risk.

Early stage professionals rarely provide those assurances. The work takes longer than expected. The results are slow or lacking. Things can be rough around the edges and the deliverables unpredictable. More coaching and direction are required but it's impossible to know how much. These are natural parts of learning, but employers view them as additional costs.

They want experience. They want an expert. The expert's work is fluid and deliberate. Mistakes are infrequent and their work is polished. Predictability provides comfort to employers. We know this, but from a hiring manager's perspective, hiring the expert is less risky and more efficient.

The question is then, where does that leave the novice when hunting for a job?

My first recommendation is to adjust expectations. Yes, there are examples of people leaving code school to start six-figure jobs. Some land a gig with great benefits or purposeful work. In each case, they are rare. While we would all love a high paying job, months of vacation, great work-life balance, and fulfilling work, that is far from where most people start.

The first tech job I (Tim) got was at a small agency building e-commerce sites. I was paid less than any developer should be paid and worked in a room packed with other engineers. We were elbow to elbow, each sharing folding tables. The room was hot, had a very distinct odor and at times felt claustrophobic. But it was an opportunity. I got to wake up every day and code. I got to learn new things and see my skills progress forward. At the time, my life was simple enough to work under those terms. In exchange, I had the opportunity to gain experience.

Because I took that chance, the next opportunity came about easier and at a better rate. Each opportunity builds upon the last. Early on, the most important thing we can do is take action. It's unlikely your first opportunity will be the best, but you can't build momentum without movement.

12

MAKING MONEY
There are multiple options to make money in tech. None are perfect.

The first paycheck I (Morgan) ever earned in technology wasn't glamorous. I didn't have a job title or formal training. It was for a company with no connection to the broader tech industry. Most of my job consisted of creating physical order fulfillment and tracking systems, not software. The day to day looked a lot like organizing a warehouse.

I started maintaining an e-commerce website on the side. Then I began building basic landing pages, also on the side. No CSS. No JavaScript. A lot of Adobe Dreamweaver. It would be years before my official role involved software engineering.

There are multiple options to make money in tech. None are perfect. Whatever it looks like for you, get going. You can and likely will change course later—and remember that the ways in which you are compensated may vary along the way.

TOTAL COMPENSATION

There is more to consider about making money than just your

paycheck. Several factors impact the compensation you receive and can make or break the economics of your work. The following outline many of the most significant factors (outside of salary) that impact the final take-home package:

- Equity / Stock options
- Health insurance
- Vacation & time off
- Retirement
- 401k match
- Commission
- Profit share
- Continuing education
- Perks
- Reimbursements

There are also the intangibles. These items do not have a direct, guaranteed return on investment but will impact everyday enjoyment and career opportunities presented throughout the work.

- Team members you work with.
- Autonomy within your position.
- Status inside and outside the company.
- Mission of the organization.
- Creativity to solve problems and choose projects.
- Pressure to complete work or hit goals.
- Upward mobility to outgrow current position.

If it seems like a lot, you are right. Comparing your options can be mind-numbing and overwhelming. There is no standard formula or true point of comparison. My primary suggestion is simple: If you're new to a career, solve for the current season of life. Instead of getting stuck in the future and long-term potential, identify the most compelling option for today. Understanding these variations and ways of being rewarded help inform your decision making as you decide on which work option you desire.

WORK OPTIONS

Most money-making opportunities fit into three buckets: employment, freelance, and entrepreneurship. There is not a right answer but there is one question I've found to be most helpful when deciding which path is best. Ask yourself, "How much am I able to invest in my career?"

For people early in their professional career there may be room to flex on things like benefits, salary, vacation, and balance. Others with commitments like children, aging parents, dependent loved ones, or debt may have less room to compromise. Our personal context frames our decisions.

That full-time life.
Employment is the most crowded option for making money. Most jobs and job hunters are looking for full-time employment. It provides the most consistency and predictability, which is desired by employers and workers. It can be more restrictive, but the upside is worth it for the comfort it provides.

New Story was my (Morgan) first serious opportunity as an employee in technology. I ran a software business for five years prior to joining the team and freelanced before then. It was comforting to dive deeper on fewer projects. As someone who enjoys control and autonomy, I recognized early that employment provides the least. Fortunately, stability becomes more appealing as my family grows and life becomes more complex. Released from the scattered needs of freelance and entrepreneurship, I am able to focus on fewer elements of the business and further develop more focused expertise.

Starting your own gig.

Entrepreneurship is also a popular path. With the fame of celebrity software engineers like Zuckerberg of Facebook, Basecamp's DHH, Sergey Brin at Google, and Apple's Wozniak it's no surprise this path is so appealing. Entrepreneurship involves developing an idea, selling to customers, and operating a business. It is the most permissionless path to making money. While high risk, it provides the most flexibility and control.

I started my first business early in my career. After a year of freelancing, my cofounder and I decided to make the partnership official. In the following years, we hired staff, built out offices, and completed hundreds of projects. Starting a business provided endless changes for personal and professional growth. It was fulfilling. My network, experience, and perspective each grew tremendously. It was also hard. While most of my fondest memories are tied to this season, so are my greatest mistakes and failures.

The best illustration for the risk of entrepreneurship is from the winter of 2017. One of my clients underwent massive layoffs. As part of their "strategic shift," they left us with unpaid invoices, accrued expenses, and contracts worth over $600,000. The hit was devastating and the ensuing months were even worse. At any moment, the team members we hired and freelancers we contracted could leave without consequence. As cofounders, things were not so simple.

The business survived but hundreds of thousands of dollars were lost in the process. A mentor once shared, "Nobody wants a ceiling on their potential, but few people are comfortable without a floor."

Entrepreneurship has the potential of more financial upside than freelancing and most jobs, but it comes with its risks. The momentum a business creates can also incur loss.

Life of the flexible and freelancing.

Freelancing is somewhere in between employment and entrepreneurship. It doesn't provide the same stability as employment but has more flexibility. It's less complex than entrepreneurship, though the financial upside is more limited. Fortunately, a season of freelancing can be one of the fastest ways to build a large portfolio of paid work.

Before I committed to building a business, most of my work was freelance. I've built websites for family, friends, and businesses of all sizes. From a homemade baby food recipe app to a network for project attorneys, freelancing provides a wide variety. Most freelancers struggle to balance creativity with operational needs, but it's hard to dispute the flexibility of freelancing.

The freelance market is filled with jobs. There are solo opportunities and teams needing additional support. Some engagements last weeks while others might last months or years. The dynamic nature of freelance software projects makes it my top recommendation for early stage professionals looking to make money in tech. Examples of simple, clear freelance might include: A brochure site for a friend, an internal tool or integration for a company without developers, or handling random support tickets for a small tech team.

Another friend of mine pushed my thinking even further when it comes to making money. She was adamant about "making the job you want." She graduated from an immersive product design course focused on User Experience and User Interface design. Years later, through a series of promotions, she became a Senior User Experience Designer at Home Depot. She tells it best, "There was a time I wanted to expand the scope of my current position. Outside of my daily work I was interested in animation, which felt like it could be the growth opportunity I was seeking. I hunkered down and worked to learn digital animation. I spent extra time educating myself and sharpening my skills. I am working towards making myself the animation expert on the team. Ongoing education keeps you sharp and competitive. It also helps you stay excited and engaged in your career."

Whichever path seems most appealing, remember you have options. Most of the money-making opportunities in tech don't have a corresponding job description posted on a webpage. It's important to start somewhere. More is within your control than you might think.

13

APPLYING TO JOBS
A business is no more than a group of imperfect people.

No one has it all figured it out. Taxes, health insurance, and dry cleaning are symbols of adulthood. Added responsibilities make us feel like adults but I have never met an honest person without insecurities about "doing it right." Adulting is no joke. Behind our polished personas are average people making it up as we go.

The job hunt reflects this reality. While the nature of job hunting forces you to confront imposter syndrome over and over again, the people on the other side of the table are likely combating doubts and uncertainties of their own. It may appear you are interviewing with businesses, but a business is no more than a group of imperfect people. Behind every decision is a person with doubts and insecurities. You are not alone.

PLAYING THE NUMBERS

Job descriptions are flawed.
Most job descriptions are not written by the person who has actually done the work. More commonly, job listings are borrowed from

random places online, modified templates, or outdated versions used for previous roles. A quick google search of "job description for" reveals many of the top options. It is unrealistic to assume the requirements and expectations were carefully crafted, representative examples of the exact role. It is possible, but it is hardly the norm.

I realized this early in my career. I was reading a job description for a Senior Ruby on Rails Developer. Skimming down to the requirements section, I noticed the employer wanted eight years using the language. At the time, eight years of experience was impossible considering Ruby on Rails was only six years old. David Hanamier Hanson, the language's founder, would not have qualified for the position. Another example is when Apple announced the Swift language. Companies realized the value and wanted to start migrating their codebases, so they put up job descriptions. Many of those descriptions required two or more years of experience, for a tool that was released earlier that day! These were more than innocent typos. Today, the tech industry is flooded with similar examples. Details as fundamental as years of programming experience and similar "requirements" are not as precise as you would expect.

Why does this matter as you hunt for your first tech job? Job descriptions are flawed. They are written and published by regular people. They are subject to the same biases, imperfections, and misconceptions as anyone. Do not expect perfect alignment.

The most thorough job descriptions I've written were at New Story. There, weeks are spent evaluating a role's requirements. I appreciate the care and desire for clarity it represents, but frankly, this is my least favorite part of a maturing organization. Despite how hard we try, we still never really know. Most job descriptions, especially when not replacing an existing role, are some mix of needs, desires, assumptions, and premature anticipation. It's slightly better than guessing.

When a candidate is hired, their experience and what they will actually do rarely looks exactly like the original job description. While the candidate is hired for their skillset, they are also hired for who they are and what they bring to the table. A few years ago, I put a job role up for a senior software engineer. In my mind, I was expecting something around five or more years of experience. We ended up hiring someone straight from code school and instead chose to modify the title attached to the role. Why? Because we believe what the individual lacked in years of experience, they made up for in drive and passion for the role. They ended up being one of the best hires we've ever made.

You need a spreadsheet.

If you don't need a spreadsheet to track the jobs you've applied to and where you are in the process, you probably haven't done enough. While helpful, the spreadsheet is not the important part. It's about quantity. The amount of activity should exceed your ability to remember it all, otherwise it's unlikely you are giving yourself a fighting chance.

To download an application tracking spreadsheet,
visit codeschoolbook.com.

Paige Niedringhaus, a code school graduate who's now a Senior Engineer at the Home Depot, was generous enough to share her experience in the application process. "I applied to close to 70 job applications, got rejections from around 10, got phone screens from two and got one in-person interview. I know many who applied to two to three times the amount of job openings I did. At that point in my career (just starting out), I couldn't afford to be too picky even if I felt mediocre about meeting the requirements or about the company's product. If it was a software development job, I applied for it. Once that first job is under your belt and on your resume, getting the next job is infinitely easier. Work experience signals to employers and recruiters you can do the job they're looking to hire for."

All applications are not equal.

While there are many outlets in which you can apply to a new job, they are not all equal. For instance, clicking "apply" on Indeed is not the same as a formal application on a company's website. Because large job boards make it so easy to apply, employers can have a hard time relying on quality candidates and are left to filter through a lot of noise. This makes you one of many in an already large pool. Your goal is distinction, sticking out is how to self-select. Wherever possible, find the role you are interested in on the company's site and apply directly.

Mile wide and mile deep.

When applying for jobs, you should go broad. Work to fill out your spreadsheet, but remember to also dive deep into a few key roles. Of all your job applications, pick 20 to 30 percent of the roles to focus on more intently. Spend extra time navigating social channels and engaging their content. Connect with people at the companies wherever they frequent online. Build unique portfolio pieces that reflect interest in their organization. Film custom introduction videos and send them to team members at the business. The combination of broad and deep ensures you've invested enough attention.

Adrianna Valdivia, a former team member and non-traditionally trained engineer, did this during her application process at Polar Notion. When asked to share a piece from her portfolio, she presented a hangman style guessing game. Instead of using a traditional stick figure, she used Freddy, the mascot of popular email marketing company MailChimp. This artifact resonated well with our team, as serious MailChimp fans. I'm sure it would have connected with interviewers at other companies around town, too. In hindsight, it probably didn't have to be company specific to be effective. Enriching portfolio pieces with relevant flares from the local community can add newness and nostalgia.

Kevin Meldau provides another great example. In 2018, we had an open position for a software apprenticeship. Dozens of applications flooded in from all over. The average candidate submitted an application. A few would follow up via email. Kevin went so far as to email, LinkedIn message, engage social content, and more. Looking at a pile of resumes, his became harder and harder to ignore. There is a delicate line between persistence and pestering, but most people fail to come close. Kevin's approach led me to at least consider an in-person conversation, which is where his thoughtful demeanor ultimately sealed the deal.

I'm providing a third example to spark your creativity and reinforce how effective it is to dive deep and invest heavily in a handful of job opportunities. It was the middle of 2020, six months into a global quarantine. Our team was working remotely, but I drove into the office to have a peaceful place to work. I heard a knock on the door. I was accustomed to frequent visits delivering packages, but this time was different. A well-dressed gentleman stood at the door with an envelope. Clearly, he wasn't from Amazon or Fedex. Greeting him at the door he called me by name and said, "Hey Morgan, I'm Tyler. Nice to meet you. I'm applying for the Executive Assistant position. I wanted to stop by and say thank you for considering my application. Would you please make sure this gets to the hiring manager for the position? Have a great day." He handed over the envelope and walked away.

A few things surprised me from this moment. First, there was no guarantee anyone would be at our office but he went for it. Second, he recognized me on-sight, which given my nonexistent celebrity-status meant he likely was prepared to encounter any number of team members. Third, inside the envelope was a handcrafted pop-up card customized for our iconic 3D-printed home project. The entire experience left me impressed and eager to pass his information along to more of our team.

14

ALIGNING WITH EMPLOYER EXPECTATIONS
Great culture isn't about ping-pong or beer Fridays.

Human behavior can be summarized in one word: incentives. Properly incentivized, we've watched people eat strange insects, race around the world on scavenger hunts, and starve themselves alone in the woods. Incentives drive people's behavior.

Employers are people too. Despite interviews, questionnaires, references, and assessments, employers are just trying to understand three things. Regardless of how they pursue their answers, employers want to know:

- Can you do the work?
- Do you want to do the work?
- Will you fit well with the team?

I refer to these as the three Cs for hiring. There are all kinds of resources out there that reference various "three Cs for hiring" but I'll point to mine as capability, curiosity, and culture.

Capability: *Can you do the work?* This underlying question speaks to skills, abilities, and experience. These factors come together to increase an employer's confidence in you to meet their needs. The best applicants make it clear they are capable without much poking or prodding.

Curiosity: *Do you want to do the work?* Believe it or not, I'm quite an accomplished lawn mower. Since middle school, I've been paid to cut grass. While I'm capable of doing the work, I am comfortable never touching a lawn mower again. Knowing humans work best when personal interest is involved, employers work to ensure team members would be engaged. Many applicants are more interested in the money than the work itself.

Culture: *Will you fit well with the team?* Google conducted a massive research study about effective teams.[1] After studying thousands of teams, they found no clear correlations that make a high performing team. It wasn't until they started looking at psychological safety that they found a pattern. Teams willing to take risks without feeling insecure or embarrassed outperform the rest.

A great culture is not about ping-pong or beer Fridays. It's about fostering a safe place for ideas and conversation to flow without fear or shame. Employers and hiring managers want to ensure the new hire will not detract from their staff's experience.

Sharing directly from Google's research, these are the five key dynamics that set successful teams apart:

- Psychological safety: Can we take risks on this team without feeling insecure or embarrassed?
- Dependability: Can we count on each other to do high quality work on time?
- Structure and clarity: Are goals, roles, and execution plans on our team clear?
- Meaning of work: Are we working on something that is personally important for each of us?

- Impact of work: Do we fundamentally believe that the work we're doing matters?

While employers are asking these three questions, culture (with psychological safety) has been proven to be the most important contributor to a high performing team. What does that mean for you? Explore ways to model and clearly represent that in the hiring process.

People rarely make the best choice. They choose what they understand the best. How might you use the interviews, reference checks, and assessments of getting a job to reinforce the reality that you are capable, curious, and fit for their culture?

HACKING THE INTERVIEW
The interview process is a perfect example of how many small investments compound into long-term success.

I t's rare to get an interview. In the hiring process, most people submit applications to dozens of organizations to get a single interview. Given how much effort is required to get facetime with employers, it still surprises me how many candidates miss the basics.

Getting an interview is hard but standing out during the interview process is not. Technical needs are a small part of successful interviewing. It is a low percentage of interviewees who do any one of the following. I have a strong conviction that anyone doing all of them is impossible to ignore. "Death by a thousand cuts" is a common expression when describing failure due to a lot of problems or issues. I prefer "victory by a thousand wins". The interview process is a perfect example of how many, small investments compound into long-term success. This list is accessible to anyone. It's also applicable across many positions and nearly any company. Without knowing the specific requirements of a role, the following suggestions provide the easiest way to excel while interviewing and increase your odds of getting the job.

Follow up.
Each year I speak to dozens of code schools and bootcamps. In every presentation I make a point to share my contact details and email, and I directly invite people to reach out. I make it clear I'd help without cost and I'm hiring code school graduates. Thousands of people have heard the same prompt, every event. But even so, only a fraction ever actually follow up.

A survey from global staffing agency Robert Half International found that after simply sending a job application, 81 percent of 1,000 hiring managers want to receive a follow-up message within two weeks.[1] Following up after an interview is even more critical. Companies will immediately disqualify candidates who don't follow up. A helpful rule of thumb: follow up with every person, every time, within 24 hours. Great follow-up messages include personalized gratitude, a brief summary, and clear next steps. Consistent, thoughtful follow-ups show an employer the role is important to you.

Even if you don't hear back, follow-ups are essential. No reply means nothing.

The best follow-up I've received involved handwritten notes for each team member they interviewed with and a package of cookies that arrived once they'd left the office. Few interviewees have created as much positive, interoffice chatter and it cost less than $20.

Invest in personal appearance.
My favorite high school teacher had an expression that served me well on countless occasions: "Dress for the job you want, not the job you have." For physical interviews, investing in personal appearance involves clean, professional attire. The company may allow employees to wear T-shirts and flip flops, but you're not an employee, yet.

Digital interviews have enough disadvantages that appearance matters even more. Along with attire, the following checklist provides a stellar appearance.

- *Optimize your lighting.* More light on your face, less on your background.
- *Adjust your camera.* The height should be eye level and your face should cover most of the screen.
- *Simplify your background.* Remove distractions and visual noise.
- *Improve audio quality.* Moving close to the camera goes a long way, as well as muting when you're not talking.

The details and attentiveness of your appearance are subtle details that stand out. As more and more of our business lives are lived in virtual meetings, it's clear who puts in a little effort.

Come prepared.
Coming prepared means showing up early and planning prior to arrival. During the interview process, you will be given the chance to ask questions. Have something ready. While some questions are better than others, the only wrong response is having nothing to ask.

Another way to come prepared is by researching the people within the company. If the company is large, start with those you're likely to meet. If the company is less than 10 people, there is no harm in familiarizing yourself with everyone. Calling people by name is an easy way to make an impression.

LinkedIn, Glassdoor, Twitter, and Google are powerful tools to help you prepare. In a matter of minutes, you can identify the founders, key leadership, customer reviews, and public sentiment. Read through popular posts on social media and recent news articles. The facts and insight you uncover will allow you to speak intelligently about the company and avoid questions you could easily have looked up yourself.

Play to your strengths.
There is little you can do to disguise weaknesses or gaps in your resume. That is natural. Fortunately, you can amplify your strengths. Acknowledge your greatest strengths and ensure hiring managers recognize them too. While there may be a fine line between confidence and arrogance, it doesn't serve you well if employers never see where you shine.

Clifton Strengths personal assessment, formerly StrengthsFinder, provides an inexpensive way to understand more about yourself and help you represent it well to potential employers. The entire assessment operates from the belief that we stand more to gain from leaning into our strengths than "leveling up" our weaknesses.

In general, researching a few personality assessments can help you learn more about yourself and prepare for interviews. While every company uses different assessments, the patterns and learnings can make you more self aware.

Below are some of the most popular tests. While the accuracy or relevance of each test is widely debated, it's helpful to find one takeaway from each.

- Myers-Briggs
- Enneagram
- RightPath
- StrengthsFinder 2.0
- DiSC

To view Morgan and Tim's results and learn more about the above personality assessments, visit codeschoolbook.com.

My final suggestion to increase your odds of success while interviewing is, "get creative." Those who think outside the box are regularly rewarded, especially when done thoughtfully. In life there are laws and there are rules. Breaking laws is bad. Breaking rules or rethinking conventional thinking, however, can unlock new opportunities. As you embrace life as an engineer, engineering your desired outcome during interviewing is a healthy thought experiment.

On the Job

PLANNING A SUCCESSFUL CAREER
Great athletes, for example, will not necessarily make great coaches.

F inding a job is hard, but it is the beginning, not the end, of career challenges. Just like it's important to define success while searching for a job that aligns with your expectations, determining your intentions within a company are also important. How do you define success in your new role?

- Is there a path to promotion?
- Are you hoping to become a manager?
- Do you want to lead a team?
- Will you own a project?

As you establish yourself within an organization, you'll find that some opportunities are obvious while others seem obscure. Large, legacy businesses have a clear hierarchy and progression path. Within smaller businesses and startups, growth might be undefined. Knowing what you want will allow you to look for opportunities and ask better questions of superiors.

There are three paths within an organization: Individual Contributor, People Manager, and Principle. While some require more experience than others, they are not necessarily linear. Each organization views the paths differently.

In recent decades, more companies are creating different tracks for Individual Contributors and People Managers. Much of these changes are a result of the Peter Principle. The Peter Principle states that, if you perform well in your current job, you will likely be promoted to the next level of your organization's hierarchy until eventually you are promoted into a role you can't perform well.[1] Simply put, people are often promoted into incompetence.

Promoting people into a new position based on success in their former role decreases the odds they'll be successful at the next position. Excellent individual performance is different from effective leadership. Great athletes, for example, will not necessarily make great coaches.

In an engineering organization, the individual contributors are those who spend most of their time writing code and executing projects. In design firms, individual contributors are pushing pixels and creating the deliverables. As a career progresses for the individual contributor, challenges, accountability, and judgement are expected to increase but it doesn't involve the need to manage a team of people.

In a popular article by Paul Graham, he discusses the difference between a Maker and a Manager schedule. The Maker needs time to focus and produce work themselves. The Manager context switches and produces work through coaching others. The Manager spends more time in meetings, emailing, and developing others. As you start your career, you don't need to know exactly which path you want to end up on, but it's good to understand the differences of each path, so as you grow you can ask the right questions.

THE MAKER PATH

The Maker Path is where most people will start and spend the first few years of their career.[2] In this path, you'll be focused on increasing your skills and getting better at your craft. This path generally starts at a "junior" or "support" role and then moves up the ladder from there. If we take an engineer role, a typical ladder might look like the following:

- Support Engineer (Junior)
- Software Engineer
- Senior Software Engineer
- Staff Engineer
- Principle Engineer

This path is not one where you will always be pounding out code or pushing pixels in Photoshop, but rather a path where you will be focused on making things. This may include being the lead on a project and ensuring others are holding their weight. You might be responsible for reviewing code or designs and giving feedback on ways to help your co-workers improve on their skills. While you will not be managing someone else, this type of role includes mentoring others on the same path.

Another piece to understand about the Maker Path is that your days will be structured around long blocks of time where you can go heads down on a project. You'll do less context switching and often be focused on a single project for a longer period of time. You'll spend less time in meetings and need to worry less about your calendar and email. Often, your manager will even work to protect your time so you are able to focus and spend more time doing what you are good at.

THE MANAGER PATH

The Manager Path is for those who want to spend more time managing people and systems then being heads down on a single task. While this role is often defined by personality, it also leans heavier towards extroverts and those who enjoy context switching. A manager will spend their days managing people and making decisions about where the projects and products will head next. Again, the ladder for this role will start at a similar place as a maker, but will usually split off around the senior level. For instance, a typical manager ladder might look like the following:

- Support Engineer (Junior)
- Software Engineer
- Senior Software Engineer
- Engineering Manager
- Senior Engineering Manager

As you can see, managers will spend the first portion of their career as a maker and will generally focus on getting better at their craft. What will start to set them apart is they will be someone who starts to enjoy the mentorship side or the project ownerships side more than the making side. They will often volunteer to run meetings and assist with higher level decisions. Someone who is more attuned to being a maker will be just as happy to stay heads down and avoid such meetings.

Which Will You Be?
Whether you become a principle engineer or a senior engineering manager, you'll likely have similar benefits, salary, and level of responsibility. The difference will be the items that you are responsible for and how you will spend your time.

The path to get to each level looks different at different companies. When you start at a new company, it's not a bad idea to ask for some clarity around their ladder structure so that you can start to think long term about your time there. Some companies have a very rigid structure with specific ways to level up. For instance, at my first company, everyone started as a junior engineer. You then had to be at the company for a minimum of six months before being able to take a test. If you passed the test, you were moved to a software engineer position with a small bump in pay. Then, you had to wait at least two years from that promotion before you could test into a senior role. Currently, this seems to be less common as it's more based around merit and growth. In recent years, I've promoted someone in less than a year to a more senior position, solely based on how hard they worked, how well they did when given ownership of a project, and how they worked with their peers.

Brenton Strine, software engineer and founder of Refcode, says, "Even a few years into my career I still thought of programming as the act of actually typing code. Now I think of programming as conversations, whiteboards, tracking down requirements, and reading through existing code."

In the end, your goal is to work hard, improve your skills and continue to learn what makes you happiest as you settle in for a long career in this field. Whether you decide you want to be a maker or a manager, both paths are attainable.

17

A DAY IN THE LIFE
Software engineering is like running—you practice how you play.

Nearly every team member I ran cross country with in high school still runs over a decade later. No one who played on the football team still plays football. Why?

In football, the training and conditioning culminate under the Friday night lights. The practice, weight lifting, and drills consume their training but it's vastly different from the performance.

Running is the opposite. Other than stretching and drills that bookend practice, most of our preparation for cross country races involved running. Running during the week to practice and running during the performance. The pace was more intense and there were more spectators, but one closely mirrored the other.

Software engineering is like running. It's a sport where you practice how you compete. Whether you're learning to code, tinkering with a side project, or working professionally, it involves the same underlying skills. The behaviors and disciplines transfer instinctively. If you enjoy critical thinking, problem solving, and learning involved in coding bootcamp, there is a good chance you'll enjoy doing it as a job. Conversely, if every moment is agonizing or dull, graduating into a career of similar work will likely just compound the distaste.

The daily life of a software engineer involves writing code, googling solutions, and reacting to error messages. It's a constant pursuit to improve, grow, and refine. If we're lucky, the challenges get bigger, the technology changes, and our skill increases. The underlying elements are largely the same.

As we think about the ever-advancing nature of a career in tech, Game Theory provides helpful context. There are two types of games: finite and infinite. Finite games are ones in which the players, moves, rules, and outcomes are spelled out. Success is clearly defined. There is an established start and agreed upon ending. One possible outcome. Sporting events and competitions are finite games.

Software engineering is an infinite game. Infinite games differ in almost every way from their fixed counterparts. There is a seemingly moving target that's constantly evolving. Problems are complex and nearly every solution can be improved upon somewhere along the way. This tension both drives the industry forward and provides a source of frustration. If it seems hard, that's because it is.

TOOLS

Along with the continuous cadence of problem solving, tools are another main staple of an engineer's daily life. Engineers love tools. Few careers interact with as many tools as software engineering. If you look long enough, you can find a widget or tool for almost anything.

There are tools to perform engineering tasks, others to collaborate, and many more for communication. The most common engineering tools are the web browser, text editor, and command line. The tools of the trade, these are required to complete nearly any work in software. There are thousands of examples of each type and even more locations online with engineers debating which is best. For as long as there have been opinions, there have been people who disagree. These are a few you will encounter almost immediately:

- Chrome vs. Safari
- Github vs. Bitbucket
- Trello vs. Asana
- Jira vs. Basecamp
- Sublime Text vs. Textmate
- Ruby vs. Python
- React vs. Angular

In reality, I have found the differences aren't nearly as important as the output they facilitate. Great teams and skilled professionals can produce great results in spite of inferior tools. There is no tool that can compensate for a poor team or a careless professional. I've found one question to be the most help when selecting a tool: Does it allow me to consistently deliver greater work with less friction?

In the end, no two developers' experience is the same. The type of company you pursue, the size of that business, and the composition of their teams dictate much of their experience. Inconsistency is the only consistent. That's why adaptability is so important. Not only is technology evolving too quickly to expect too clear a standard, context varies too. Engineers must stay flexible and focus on solving problems. In nearly everything else, your mileage will vary.

18

SOLVING PROBLEMS
How many team members does it take to change a light bulb?

A few years after starting our first business, our growing team moved to a new office. We built out the space with beautiful accents, stylish furniture, and unique artwork. For the first time, it was a place of our own. Accompanying the excitement, we realized how much work went into keeping a space maintained. Things seemed to be in constant need of repair.

After returning from an off-site meeting one afternoon, a team member pulled me aside. They proceeded to explain that our conference room light had not been working all day. I waited for further clarification but none came. There was a single conference room for the entire team. Six hours after identifying a problem and "the light won't come on" was the extent of their investigation. During this time, other meetings occurred and a number of people chose to tolerate the same inconvenience.

Choking back frustration, I walked into the conference room and flipped the light switch. Sure enough, the room remained in darkness. I turned on the flashlight from my phone and inspected the lightbulb a few inches overhead. The bulb was busted.

Walking eight feet to our storage cabinet, I pulled out a new light bulb, stood on a chair, removed the expired bulb, and attached the replacement. In less time than it took for the team member to relay the problem, it was diagnosed and solved. A burnt-out lightbulb.

People are skilled at recognizing problems. We see them, talk about them, and debate them. It's easy and requires little effort. Most people stop at problem identification.

Great technologists are solution oriented. Solution-oriented individuals push to understand the underlying problem and craft thoughtful solutions. Regardless of the size of a problem, making an effort to diagnose and propose solutions goes a long way. The practice improves our personal problem-solving abilities and makes team members invaluable in the workplace.

Ben was Polar Notion's first apprentice. He was solution oriented from his first week on the job. When he encountered an issue, he started formulating his solution almost immediately. Then, he would present it to me and we would refine it together.

Sometimes his approach was correct. Other times, it was completely wrong. Whether right, wrong, or somewhere in between, the important habit was making the attempt. With each attempt and revision, the quality of his solutions improved. It was through taking action that he increased his practice and perfected his problem-solving skills.

Similarly, team members who proactively solve problems set themselves apart from their peers. In medicine, doctors participate in a process called differential diagnosis, where they differentiate between two or more conditions that could be behind a person's symptoms. Ideas and suggestions are bounced around until an appropriate answer seems to fit. If they withhold their ideas until after their confidence increases, the process would take longer and their solution would lack the benefit of other experienced insight. The mental thrashing is a valuable part of arriving at the best outcome.

19

MANAGING TRADEOFFS
There are problems to solve and tensions to manage.

A t a young age, most students receive tests. Each question on the test has one correct answer. During quizzes and exams, we earn some percentage of a "perfect score". An instructor defines "success" and our answers as either wrong or right. There is no gray area or middle ground. There are wrong answers but only one right option.

There are few areas in life when this binary thinking serves us well. In most instances, especially software engineering, things function differently.

Rather than weighing right versus wrong, software engineers are confronted with endless tradeoffs. When faced with many acceptable answers, software engineers understand that some are better than others. We understand there are many wrong answers, but some are worse than others. There is no right answer.

When we realize there isn't a right answer, it can be immobilizing. Without certainty, how do we know what we are doing is right? Acknowledging tensions and working within them is both the greatest challenge of engineers and our main value. We are expected to create and solve problems amidst uncertainty, not without it.

Andy Stanley, an excellent teacher and author on leadership and work, says, "There are problems to solve and tensions to manage." Simply put, things that are broken should be fixed. Tensions, by contrast, must be managed. Issues emerge in our work when we don't understand the difference between the two.

Consider the analogy of a car. While in motion, brakes apply tremendous friction which allow the car to stop. The tension between the accelerator and the gas pedal is essential for the car to function. Trying to remove that tension diminishes the overall value of the vehicle. Now, imagine the headlights didn't work. This is a problem. Driving at night would be very dangerous. You could manage the problem by only driving during the day, but what would happen if an emergency arose late one night? The broken headlights don't need to be managed; they need to be fixed.

In technology, common tradeoffs include time, features and budget. These tradeoffs are complex because their forces are interconnected. They do not exist independent of one another, nor do they behave consistently. An increase in one area could produce a negative impact on another. Conversely, an increase in one might increase another— or both.

Let's explain how this plays out when building an app. Adding features to the software product will impact the completion date. You could reduce time by adding more team members, but that would increase the budget. You could also decrease the time by reducing the number of features. Which is the right answer? It depends.

It's nearly impossible to make an informed, thoughtful decision without more context. To gain perspective, I like to consider the question, "What are you optimizing for?" This forces a conversation around priorities. As priorities are better understood, decision making can be improved.

Within every choice we make is the inherent opting-out of other options. This concept accompanies tradeoffs and is known as opportunity costs. Opportunity costs capture the missed opportunity among the options we didn't choose. Given limited resources, we cannot do it all. For example, a friend invites you to dinner. If you accept their invitation, there is an unspoken understanding that you won't be attending other dinners happening during the same time. When saying "yes" to anything, what are you saying "no" to?

Blanca Garcia, entrepreneur and friend says, "Clarify the life you want to live." As you consider a new career path, define what you're optimizing for. Are you looking for the ability to buy nice stuff, flexibility, upward mobility or influence, etc.? What are you looking for? What is driving you? What causes you to endure? Does this path give you the greatest chance of achieving it within the time and effort you have?"

Misunderstanding tradeoffs is a common source of frustration between technical and nontechnical teams. Oversimplifying tensions can lead some to treat them as solvable problems. In tech, pregnancy is used as one of the most common comparisons when discussing the tradeoff of effort and time. A woman is pregnant for about nine months to deliver one child. However, nine women can't work together and "get the job done" in one month. The belief that you can put more people on a problem is popular, but rarely as simple. Some things just take time.

20

EXCELLENCE, NOT PERFECTION
How we work stands out to others more than what we work on.

Reference checks are the most informative part of the interview process. They cut through the theatrics. References share real world insight into how this person has behaved on the job. Two of my favorite questions for references are "Is this person in the top 1 percent, 5 percent, or 10 percent of the professionals you know?" and "How would you describe this person in one word?"

Once they respond, I ask them to elaborate on why. Over the years, I've gotten hundreds of responses. References rarely remark on programming skill. I hear little about code writing ability or features built. At a certain point, those things are just expected. Instead, references speak of work ethic, team work, and diligence. While some characteristics are ingrained, many can be built. And in doing so your overall perceived value will increase.

YOUR WORKPLACE VALUE

How we work stands out to others more than what we work on. The most remarkable professionals are those who focus, prepare, respond, follow through, and follow up.

Focus.

Focus in the workplace isn't as common as you might expect. From the C-suite to the most junior employee, distractions are everywhere. If organizations and teams aren't careful, it chips away at productivity and an individual's ability to do meaningful work.

Software engineers excel when they prioritize focus. Whether they have to advocate publicly or develop habits privately, engineers need uninterrupted periods to dive deep into their work. Multitasking undermines an engineer's productivity.

Time blocking has been the largest contributor to my ability to focus. Each week, I set aside at least five blocks of time. Each block is two to three hours and is guarded from interruption. By maintaining these blocks consistently, team members begin to anticipate this time and work around them. Time blocks are the simplest way to signal to the team and ensure time is set aside each week to push my highest value projects forward.

I didn't start with five blocks nor was each block three hours. It began with smaller chunks of time at less frequent intervals and increased over time. As managers and team members witness the effectiveness of these periods of focus, they will not only respect them but will often add similar holds on their own calendar.

Time blocking isn't the only way to increase focus. These are a few behaviors that lend themselves to improved focus:

- Scheduling recurring events each day for two to four hours.
- Listing your ideal work conditions.
- Listing your weakest work conditions.
- Creating agendas for time blocks each week.

- Identifying common distractions in your typical work environments.
- Identifying "amplifiers" in your typical work environments.

Prepare.

I (Morgan) love pizza. When given the choice of what to eat for dinner, my suggestion is almost always the same. Early in our marriage, my wife would travel in three-day stretches every few weeks. Without fail, I would eat pizza for at least three meals. I could order a large and eat leftovers for the following lunch and dinner. I was also fond of picking separate locations and ordering from a different restaurant each night.

When friends get together, pizza is our typical meal too. It's not because my friends are as enthused about pizza. It's because, when we poll each other for suggestions, I'm prepared to answer. I know what I want. And because of this, I'm usually the first to respond and present the strongest perspective. When groups come together and seek consensus, there is a lot of power in simply speaking up first. Being prepared provides disproportionate opportunity.

Similar dynamics are true at work as well. Group norms are often set through consensus, not by considering what's best. For this reason, things like productivity or effectiveness become diluted as teams grow. But you can protect your, and potentially your team's, productivity.

Preparation is a powerful tool in doing so. The most prepared person in a group often sets the tone and benefits from a first mover advantage when asked to speak. A popular comparison is that of thermostats and thermometers. Thermometers reflect the temperature in the room. Thermostats change the temperature. One reflects the climate while the other alters it. Given the rush and chaos of many organizations, teams and their employees fall into weak habits around preparation. Investing time and energy into thoughtful opinions before they're needed is the simplest way to lead and change the temperature.

Practically, a disciplined meeting culture is one of the most common ways to exercise preparedness. It's an important part of standing out within an organization. Meetings are a company's largest expense. If we aren't careful, they are the main barrier to doing great work.

A little structure goes a long way. Here are suggestions to improve preparedness and limit the impact of meetings on your schedule:

- Default to shorter durations (15 minutes).
- Create the agenda before you request a meeting.
- Set a moderator/owner for the meeting.
- Set a timekeeper for the meeting.
- Send updates and details ahead of meeting.
- Avoid "bookending" meetings without breaks in-between them.
- Use the "travel time" feature on calendar events to plan for moving between meetings.
- Cancel unnecessary meetings.
- Try to "outpace" the need for a meeting using written communication.
- Ask constantly, "Which meetings can I cancel?"
- Start on time.
- End on time.

You must control your calendar or it'll work 24/7 to control you.

Respond.

The average professional gets dozens of emails and hundreds of messages per day. For many, their inbox or chats become an unwieldy mess. It is not surprising that tasks, conversations, and information get missed. Staying on top of it all is challenging and as new forms of communication infiltrate the workplace, it's only getting harder. This is part of why responsive individuals are increasingly impressive.

There is a difference between "responding" and "answering." One of the most common reasons items accumulate in our inboxes is because we do not have the answer. Either because it requires more information or we don't have time to process it, we withhold further communication until we have the answer the sender is waiting for. Instead, we should consider responding, even if we're not ready to answer. Even without an answer to someone's inquiry, a response acknowledges that you have received the request and allows you to share a brief status update about when you will know more.

Being responsive protects the time of others. The following suggestions provide thoughts on improving our responsiveness:

- Respond within 24 hours (1 business day).
- Question others' priority ("How urgent is this?").
- Communicate bottlenecks they might not notice.
- Maintain internal communication in Slack (channels not DMs).
- Maintain external communication in email.

A final technique I use is a heads up in my email footers. Each email includes the following message underneath my signature: The added communication takes a personal habit and sets better, clearer expectations with people who message me.

FYI... I use email differently. Prioritizing human interactions, I only check email twice a day and don't have notifications enabled. I do however, work to clear my inbox at least once each day.

The message includes a link to a blog post interested people can follow to read more about why I've made this a habit in my life.[1] The added communication takes a personal habit and sets better, clearer expectations with people who message me.

Follow up, again.

We all have moments where co-workers or clients ignore our emails. Imagine you and I are sitting together in a room. Assuming you had my attention, it would be rude of me to not respond when you ask a question. In the physical world, this offense is obvious and often painful. These instincts don't translate online.

Social norms are different on the internet. It is partially why the responsiveness mentioned above is so remarkable. We are inundated with messages, chats, conversations, notifications, pings and more. There is too much noise to expect others to adopt a similar standard on responsiveness. When we don't hear back, we often assume it's deliberate but there are hundreds of reasons why this may occur. Because of that, I underscore the importance of following up. In the digital world, no reply means absolutely nothing.

My personal rule on follow-up is simple: Every person, every time. People who follow up stand out. Those who follow up after meetings and requests make an impression and are more inclined to get what they want.

Here are recommendations to improve your follow-up:

- Send a follow-up email for every meeting you attend.
- Include next steps, resulting tasks, and overview.
- Recap weeks and months later on progress.

- Set a monthly-quarterly check-in cadence for mentors and advisors.
- Use follow-ups to ask for feedback.

If you need a response, these points are specifically helpful:

- Group actionable items.
- Use bullet points for tasks.
- Use a numbered list for next steps.
- Follow up at increasing intervals.

Reflect.

Whether direct or indirect, we are constantly exposed to feedback about our performance, abilities, and growth. Reflecting involves taking the time to pause, see what is working, and identify areas for improvement. If we aren't careful, weeks or months can go by without adjusting how we work.

Reflection and the resulting course corrections are a key part of getting better. Many organizations include post-mortems or retrospectives after each engineering sprint. These sessions expose flaws in the process and gaps in the team. Applying the same method to other areas of life has similar results.

I have found three areas where reflection fits naturally: daily, weekly, and annually.

Daily reflection can be fast and simple. I use five to ten minutes to reflect at the end of the day or first thing the following morning. These are questions that help with daily reflection:

- Did I complete what I expected? If yes, what contributed to this success?

- What came up that I was not expecting?
- Were there any useless meetings?
- Did I enjoy the day?

Weekly reflection is a little more involved. Twenty to thirty minutes is a fraction of the overall week but often provides enough time to reflect. Just as important as the following questions are the resulting changes or improvements to the upcoming week.

- Did I follow up with everyone I met with?
- Did I enjoy the week?
- Where was I most impactful?
- Where did I get stuck?
- What is my standing with key team members?

I dive deepest on annual reflections. A year consists of almost 9,000 hours, so it seems reasonable to spend around three of those hours looking back. The goal is not a resulting task list for the next year. The purpose is to refocus yourself around what matters most. I want each new year to include more of what matters and less of what doesn't. There is a complete breakdown of my annual reflection online at codeschoolbook.com, but the reflection includes personal exploration of the following areas. I start with a basic one to five-star rating, then review my biggest mistakes and failures for each:

- Family (spouse, children, parents, siblings)
- Relationships (friends, peers, mentors, mentees)
- Health (mental, physical)
- Profession (fulfillment, purpose, boss, team/department)

The desired culmination of these subtle tweaks is an ever-improving you. Over the years, I've had a hangup with the word "excellence." I've struggled to feel worthy of using the word excellent to describe anything I did. Early on, it all felt amateur and messy and the furthest thing from excellent. The concept of excellence felt inaccessible and somewhat elitist. But when I joined New Story, I encountered a company value that changed my perspective: "We humbly pursue excellence." Combining humility with excellence, and defining it as a pursuit caused the disparate pieces to click into place. I may not feel worthy of excellence, but I can agree it's a worthwhile pursuit. You can too.

21

TOOLS OF THE TRADE
The quality of your work will rely on your tools and of your level of comfort with them.

Prior to software engineering, web browsers were the main software I used on my computer. I remember how confusing it was early on to bounce back and forth between the browser, command line, and text editor. I was constantly typing console commands into a text file or hitting the save command within a terminal window. Not only did my understanding of the computer increase over time, but so did my dependence on its many capabilities.

During code school, you were introduced to many tools. Nothing compares to the number of tools that appear while you're on the job. There are many brands and options within each category, but the following list outlines the tools of the trade.

Beyond the tools associated with your respective programming languages, these are the types of technology you'll encounter on the job with a few examples of each:

- Web Browser (Chrome, Safari, Firefox)
- Text Editor (Sublime, VS Code, VIM)
- Notes (Notes, Evernote, Bear, Obsidian)

- Project Management (Jira, Asana, Basecamp)
- Source Code (Github, BitBucket, GitLab)
- Chat (Slack, Teams, Hipchat)
- Docs (Docs, Word, Notion)
- Calendar (iCal, Google Calendar)
- Video Call (Zoom, Hangouts, Slack)
- Dev Ops (Heroku, Digital Ocean, AWS)

I have shared this list of tools with code schools around the world. Regardless of the group, most want to know, "Should I learn these before landing a job?" An interpretation that's closer to what we're really asking is, "Will potential employers expect me to know these in order to get a job?"

The most reasonable answer is no, it is unlikely you'll be expected to know each tool. In most of these categories, there are too many options on the market to expect deep knowledge of anything in particular. Even still, if you're struggling to pick something, here are our favorites within each category at this time and place in 2021.

	Morgan	**Tim**
Web Browser	Chrome	Chrome
Text Editor	Sublime	VS Code
Notes	Obsidian	Sublime
Project Management	Asana	Jira

Source Code	Github	Github
Chat	Slack	Slack
Docs	Notion	Notion
Calendar	iCal	Fantastical
Video Call	Zoom	Zoom
Dev Ops	Heroku	Heroku

To view an updated list of these tools, visit codeschoolbook.com.

Try these out or discover your own. The quality of your work will rely on your tools and of your level of comfort with them. At this point we've already discussed imposter syndrome and mindset and how to leverage those to your advantage. The same applies when deciding which tools to engage. This is your journey. Let the ample lists of resources create inspiration, not overwhelm. This is your opportunity to begin to carve your own unique path in the coding forest. Where will you go?

The Underbelly

22

DEBATE AND DISCOURSE
Lots of opinions and ideas, but no guarantees.

R uby on Rails was the first programming language I learned. Early on, I was concerned my chosen programming language was wrong. The internet is flooded with justifications both in favor and in opposition to Rails. With so many options and constant additions, how can we be sure we have chosen correctly? The same unease emerged when I noticed a friend was using a TextMate editor and I was using Sublime Text. What does it mean when your smarter, more experienced friend makes a different choice? What about the person who uses Bitbucket instead of Github to store code and collaborate? You wonder if they know something you don't. You question if you've made the wrong choice.

This cycle continues for web browsers, version control, web hosting, domain providers and more. The pressure is compounded by public argument and endless articles debating the merits of each conceivable alternative. You think the tension subsides when you settle into a handful of tools and immerse yourself into a community. Unfortunately, disagreements rage on as super users push various configurations.

In 2016, I (Morgan) was at a conference listening to Marc Randolph, one of Netflix's founders.[1] He shared where the idea for Netflix came from and the early days of slogging it out. His talk was filled with stories about the ups and downs from their DVD-by-mail business, moments they nearly lost everything, and the insights that unlocked their eventual breakthrough. The most memorable quote from his session was, "Nobody knows anything."

The quote originated from the playwright of *The Princess Bride* in reference to Hollywood gatekeepers. Whether a movie would become a blockbuster or a bust, no one knew. The century-old industry is filled with experts, insiders, critics and theorists. No one knew, beyond a shadow of a doubt, that something would work. In 2017, $300 million was spent to produce *Justice League* which went on to lose nearly $100 million. Eighteen years earlier, *The Blair Witch Project* grossed millions with a production budget in the thousands. Nobody knows anything.

Randolph drew an easy comparison to the startup world. Lots of opinions and ideas, but no guarantees. Since hearing those words, they've rang true in nearly every area of life. The software industry is no exception. There is no tool that's perfect, nor one choice that fits everyone. The tech industry is immersed in debates that have no right answer. Absolutely, we are entitled to our opinion and it might be worth sharing, but nobody knows definitively. Rather than losing ourselves in endless debate or being lured into overconfidence, the healthiest perspectives I've witnessed acknowledge a simple truth, "It depends."

This idea plays out on both sides. First, as you begin your career, you'll make decisions around the tooling and processes that work best for you. To succeed, you need to hold your ground and also be open to learning new things. If you attend a bootcamp to learn JavaScript, then stick with that for as long as it makes sense. It's not worth second guessing yourself. Resist the urge to listen to others who claim you should have learned a different language. The unproductive contemplation will cause you to feel even more like an imposter than you already might. Your time is better spent focusing on refining your current skills, making the most of them, and looking for new ways to apply them.

Likewise, be careful how you interact with those who are still learning. One of my co-workers regularly uses an outdated text editor. It hasn't improved in years. Every time they open the application, I want to speak up and say, "Why are you still using that? You should use this instead." This open judgement of their tool choice could produce harmful insecurities, especially coming from a senior leader. What other decisions might this cause them to doubt? In reality, if it doesn't seem to impact their work and they function well, the difference is trivial.

The way you were taught or learned is not the only way. Every code school and online curriculum is different. Everyone connects with concepts and tools in unique ways. Lean into what works for you. Let others lean into what works for them. Being open and receptive to new ideas and perspectives will help make the industry better.

Lastly, if you have a strong opinion about a specific topic, the industry has a standard outlet: write a Medium article. The act of expounding upon your thoughts and ideas in long-form content is a clarifying process.

23

DIVERSITY, INCLUSION, AND EQUITY
The tech industry has done a poor job of fostering an inclusive environment.

Of all the chapters in the book, this was the most challenging to write. We don't want to get it wrong, misrepresent, or speak out of turn. We are not experts on this topic, yet we could not write a book about technology without calling attention to the elephant in the room. The tech industry has done a poor job of fostering an inclusive environment for women, people of color, and the LGBTQIA community. The authors of this book have benefited directly from our own privilege.

The growing number of authors' voices in this arena are invaluable. Many reflect the lived experiences of those negatively impacted by industry deficits. On our website, codeschoolbook.com, you can access a dynamic list of publications regarding the topics of race and inclusion that we've found helpful in our contexts. These titles include:

- *Brotopia* by Emily Chang[1]
- *So You Want to Talk About Race* by Ijeoma Oluo[2]
- *White Rage* by Carol Anderson[3]
- *Leapfrog* by Nathalie Molina Niño and Sara Grace[4]
- *I'm Still Here: Black Dignity in a World Made for Whiteness* by Austin Channing Brown[5]

Perspective from our friend Nate Washington, Chief Technology Officer at Qoins, speaks volumes to the topic: "I think for me, one of the things that I've encountered (especially in the early years of my career) is that there were many people that questioned my expertise because I a) looked different, b) didn't have a degree, and c) took a non-traditional path to becoming an engineer. They were wrong of course, but I know that countless other non-white male developers have likely experienced the same thing."

In much of this book we've discussed concepts like imposter syndrome, the experience dilemma, and preparedness. It is possible those have impacted your path. But for many, it's likely more complex than just imposter syndrome. Instead, it's more likely you feel the tension of a field that was not built to include everyone at the keyboard. You may have an experience dilemma or plenty of experience, yet are faced with the dilemma of short-sighted employers and biased recruiters. Everyone is not starting at the same start line.

Another friend of ours and former code school graduate shared her experience with us when working at a tech company as one of few women.

"I wouldn't say I had any outwardly negative experiences being the only woman there. Everyone was nice and respectful to me, and I enjoyed collaborating with my peers. Admittedly, I thought it was kind of cool that I was the only woman in some ways, like I was pioneering a culture change within the company. Most times, I wasn't really conscious of my gender because we would generally work in small groups. But during weekly company meetings or lunches when the whole company gathered, it was definitely apparent that I was a woman in a group of fifteen dudes. It can be hard to speak up in those situations, particularly while also being new and just trying to adjust to a new work environment.

"And then COVID hit. Overnight, we all began working remotely. I did feel isolated, but we were all physically isolated at the time so I chalked up a lot of my feelings to that and the initial stresses of COVID lockdown. Still, everyone was very kind to me, but I came to dread weekly company Zoom meetings. It felt hard to connect with my coworkers in that way, and whenever conversations casually gravitated towards something "bro-ish"—for example, a Call of Duty game that some of the guys were on—I felt I had nothing to contribute to the conversation.

"It wasn't that I was the only woman to ever work at this company. Over the time that I worked there, I would hear stories about previous project managers, who were women, that briefly worked at the company. One of whom started dating a developer on the team and, from what I understand, was let go for that reason. The developer she was dating still worked there when I started. I came to understand that there really wasn't a history or blueprint at this company for a woman maintaining long term employment, which leads me to my ultimate takeaway from this experience.

"While our team was diverse in race, it was a male dominated space. Obviously, there were no women in leadership roles at this company, and I think it would have transformed my experience if there were. It would have been great to have seen an example of a woman whose perspective was valued within the company, rather than just hear stories of the brief tenures of women before me."

The conclusion here isn't pretty or polished. The problem is far from over and we have only scratched the surface. As two bearded, white men, there are subjects and contexts we are not fit to expound upon. Libraries could be filled with countless stories and experiences of the damaging effects of discrimination, misogyny, and homogeneity in technology. In many cases, by the hands of people like us.

For white males reading this message, resist the urge to get defensive. I get it. It's natural to feel attacked and disrespected, especially when matters like race and gender are mostly factors we were born into. I've felt anger, confusion, and villainized around the topic of diversity. There are moments where my success is attributed more to my privilege and whiteness than my hard work, determination, and persistence. Defending myself often feels "damned if I do, damned if I don't". And yet, I can't ignore the larger conversation. As a caring member of the community, I must work to engage thoughtfully. I've had to accept that even the best engineers are subject to bias. Bias is particularly common when we benefit from a perspective. The solution isn't to run from new insights, fresh perspectives, or contrary perspectives. Instead, great engineers lean in and work to understand before seeking to be understood.

Technology is known for creating an echo chamber. People assume their community and Google searches are defaulting to well-rounded perspectives. That's not true. The internet's tracking and curation are built to keep users engaged, which usually means reinforcing current beliefs and existing world views. In order to broaden our perspective, we need to diversify our sources. The easiest, and most actionable way I've found to broaden my understanding is to actively follow people from different backgrounds. There is an online appendix of Twitter accounts, forums, and communities where non-white voices are represented on our website, codeschoolbook.com. I encourage you to engage these sources.

For underrepresented minorities reading this, your experience is unique but you are not alone. The industry is not changing fast enough but there are communities and resources that are putting stakes in the ground to equip you for a career in technology. Almost everyone I know who has lasted in technology have found affinity groups that align with their lived experience. Navigating this industry alone can be a slog for anyone, but it's compounding for individuals outside of the dominant culture. We've curated a few to help you get started. To find an index of tech groups that amplify minority voices and people to follow or join with, visit codeschoolbook.com.

Whether or not you align with the social motives of diversity, there are practical implications as well. In Scott Page's book, *The Diversity Bonus*, he makes a definitive business case for the value of inviting diverse lived experiences into business environments.[6] In short, within the knowledge economy, diverse teams outperform their nondiverse counterparts.[7] Either way, there is no easy path forward. The road ahead for improved diversity, inclusion, and equity within technology is possible. It is hard, but work worth doing.

24

LABOR GAP
The industry is accelerating faster than the workforce is growing.

I t's no surprise. There is a labor gap in technology. I'm sure
you've seen the articles that talk about it. There are endless
assumptions, predictions, and discussions about why the gap
exists and what should be done about it. If you want a thorough
dissection, you have come to the wrong place. I'm not going to
explain why the gap exists, rather acknowledge it and share why it
should matter to you.

In many ways, the industry needs you more than you need it. A labor
shortage does not mean tons of engineering jobs are lying around.
Wouldn't that be nice. Instead, it means the industry is accelerating
faster than the workforce is growing. Unless the breadth and depth
of experienced workers increases, the demand for technical workers
will only continue to rise. As you invest in developing your skills,
take comfort in knowing most of your best opportunities are ahead
of you, not behind.

Along with acknowledging this gap, we should also acknowledge that
many low-skill jobs will likely be taken over by AI and machines in
the coming decades. What will remain are jobs requiring knowledge
work. In fact, much of the AI created will be built by folks in this
industry. It not only means the gap will continue to widen but also

ensures jobs in this field will have a long shelf life.

This is where code schools step in. They provide the accessibility to upskill into technology, start your career, and set the stage for long-term advancement. As of today, they are one of the greatest ways to close this labor gap. This means that by reading this book and taking that step, you are already doing your part to help close the gap as well as entering a new field that will give you a long, successful, and enjoyable career.

The growing gap is largely why I'm confident when encouraging early-stage engineers to pursue education and experience, even over earning a large salary. As you establish a portfolio of production-quality work, learning is more important than earning. As you gain expertise, it will likely become easier and easier to find high paying positions.

As you endure through the early days of your career, trust the process. Logging early wins, opportunities only seem to multiply over time. It's a calculated risk, but the longer the gap exists the safer of a bet it becomes.

25

EFFECTIVELY HUMAN
At our best, we are people building technology to serve people.

In the world of software, hardware, crypto-currency, AI, and endless other jargon, it's easy to get lost in the sea of cold, lifeless machines. Too often, calculated and emotionless decision-making is regularly celebrated. Without looking too far you can find examples of businesses putting profit before people and pushing unnaturally hard to maximize the mighty dollar.

This trend began during the industrial revolution. The craftsmanship economy disappeared. Humans quickly became viewed as replaceable cogs in an ever-turning engine. Seemingly overnight, the workplace changed. In recent decades, computers and robotics accelerated the trend.

As mentioned in the previous chapter, we are fast approaching a future where humans do not engage in any form of menial work. Then, the only jobs left will be those that require creativity, collaboration, and adaptability. Technologists, such as product designers, software engineers, and data scientists, will benefit greatly from this future. We also have a responsibility to retain our humanity in the process.

At our best, we are people building technology to serve people. We must interface with machines but can't lose ourselves in the process. This tension became a regular lunchtime conversation at my first company. We felt ourselves balancing the tension of efficiency and empathy. In turn, we coined the term Effectively Human. It captures the marriage of technology and humanity.

Years later, the philosophy of Effectively Human was put to the test. It was a particularly challenging season in business that I mentioned earlier. The conventional choice would have been to layoff the staff, alert our vendors they wouldn't be paid, and file bankruptcy. That was the choice our client had made that left us in this position and it made sense to follow suit. Instead, we opted to preserve jobs, repay people we owed, and work tirelessly to get the business back on course. It took two years, but everyone was eventually paid.

JUST BE NORMAL

As the world constantly changes, it only gets foggier. As technology takes over more of our lives, what it means to be Effectively Human constantly evolves. With a constant push for bigger, faster, and more, it is easy to veer off course. I've developed a few principles to hold myself accountable:

- Lead with a human-centered perspective.
- Engage in considerate debate.
- Prioritize people before process and profit.
- Seek empathy before efficiency.

Lead with a human-centered perspective.
Who are we optimizing for? Keeping the main thing front of mind,

people are the greatest prize and what we optimize for. Revolting against the industrialization of humanity, we see people as unique contributors in an ever-evolving story. The rights, privacy, and security of others come before the needs of business or our own self interest.

Engage in considerate debate.

Respectful disagreement should happen regularly. Championing the notion of "strong beliefs held loosely," few topics are off limits but conversations must remain dignified. Diverse perspectives and beliefs expose our own blind spots and drive us all toward better outcomes. We do not have to agree to coexist, but we do need to prioritize compassion and respect. We seek empathy before efficiency.

People before process and profit.

As we grow, efficiency and humanity continually square off. The fastest path is rarely the most equitable. It's not the most convenient, but our greatest work often comes from being open to unpredictability.

Seek empathy before efficiency.

Human-centered care is unlikely the fastest path forward. At times it can feel inconvenient and even frustrating when you want to move forward. But over my years of hiring and building teams, I've discovered what many already know to be true: if you want to go fast, go alone; if you want to go far, go together. The relational equity earned from seeking empathy in the workplace before (not instead of) efficiency has proved invaluable. It's a tight-rope walk and requires learning. I've also failed many times, but with this balance in mind your work character will stand out from the crowd.

The Future

26

LIFETIME LEARNER
The pace at which this field moves is equally the best and worst part.

L.P. Jacks says, "A master in the art of living draws no sharp distinction between his work and his play; his labor and his leisure; his mind and his body; his education and his recreation. He hardly knows which is which. He simply pursues his vision of excellence through whatever he is doing, and leaves others to determine whether he is working or playing. To himself, he always appears to be doing both."

Curiosity and a willingness to explore new information are key parts of continuing to grow and excel. For the most skilled, learning is a lifestyle. It's not a season, nor is it limited to a classroom.

Learning can take many forms. While some enjoy diving into a deep understanding of a small number of topics, I've enjoyed casting a wider net. Going wide instead of narrow with my learnings has exposed me to more diverse concepts and industries. That knowledge can help bring fresh perspective and insight to my current context.

My friend Brenton Strine likened lifetime learning to a game of soccer. "Continual growth has to be noted. I think a lot of people think that they can exert X amount of energy for Y years until they achieve competency, and then things will get easier. Then they'll be able to exert 1/10th the energy and just coast. But in fact, you never stop exerting that same amount of effort: you just get faster. It's the same reason that soccer is the best workout. Anyone at any fitness level can play soccer and they will get a full workout: as you get more and more fit, you don't ever get to a point where you stop breaking a sweat: you just go farther and faster with the same effort."

Being a lifetime learner is important for every aspect of your life, but as an engineer it's a requirement. The pace at which this field moves is equally the best and worst part. On one hand, it will keep you on your toes, and keep the field exciting. On the other, if you step away for a bit, you can feel left behind.

This might sound crazy, but I (Tim) can remember one point in my career when I took a few weeks off. I came back and the top choice for a JavaScript framework had changed drastically. This happened in a matter of weeks. The community made a decision and I had to either get on board or get left behind. I'm probably being a bit more dramatic than it actually was, but I remember buying an online course my first day back so I could get up to speed.

There was one morning when I was teaching at The Iron Yard and I taught my students how to use a specific tool. That afternoon, about two hours after class ended, Google announced it had bought the tool we were using and was shutting it down, effective immediately. I spent that evening learning another tool so I could teach that to my students the following day. Spoiler, that second tool is no longer around either.

My point here is not to scare you, but more to reinforce the mindset of what it looks like to be a lifelong learner. While paying attention and keeping up with the constant changes might seem daunting, it truly is one of the things that make this industry the best. Processes keep getting better. Tools keep getting more accessible. The industry keeps getting more welcoming. Trust me, it's a fun ride.

27

THE POWER OF PURPOSE

I know we like to have our cake and eat it too, but some expectations borderline on wanting the entire bakery.

For centuries, individuals performed work out of need and obligation. It was simpler. Problems existed within communities, so able-bodied people rallied together to address them. As civilizations became more advanced, so too did our expectation of work. Today, we expect much more from our jobs. Beyond merely paying the bills, we desire enjoyment, comradery, and even purpose.

Entry level candidates today want to make top-of-market compensation, collaborate with excellent people, have a positive environment, tackle meaningful challenges, and feel personally fulfilled. Wow, that's a lot to ask on day one of a new career. I know we like to "have our cake and eat it too," but some expectations borderline on wanting the entire bakery.

Whether in work, relationships, or decisions, we have to be careful pursuing an idealized option. Growing up, my friend had a healthy perspective about expectations. They would share that everyone wants a perfect fit but we miss the obvious problem: you are not perfect. We create a glamorized picture, our ideal, without recognizing that our flaws and shortcomings will likely tarnish it. I don't suggest lowering your standards. Instead, we as a community should broaden our perspective to account for the reality that we too are in process.

The more expectations you place upon yourself and your career, the harder and longer you should expect to work for them. Getting what we want from work takes time. It rarely happens quickly and doesn't appear all at once. Work is not a treasure hunt. It's more like a whirlpool.

Growing up, I remember summers in a friend's pool. It was a large, round pool but shallow enough for everyone to touch the ground with their heads above water. Most days, the water was relatively calm. From time to time however, we would all start walking in the same direction around the edges of the pool. With each lap around, the water would move faster. Eventually we'd be forced into a run. Any stray toys or floats would be pulled toward the center. If we kept at it long enough, the strength of the water became too much, pulling our feet off the bottom and spiraling our bodies inward as well.

Often, a career in tech works much the same way as that whirlpool. At first, it's nothing impressive. Each early step is heavy and hard-won. Slowly things begin to pick up speed but if we let up, the energy would dissipate immediately. Somewhere along the way, things change. It's not a particular moment as much as a subtle sensation. With enough consistent movement, the water starts shifting in our favor. The steps become less burdensome, almost easier. We begin feeling a force pushing against our back, propelling us forward.

Eventually, a strong current will form. As new desires and curiosities emerge, they are swept inward, benefiting from our previous effort and investment. When all is flowing together, it's a euphoric experience but this takes years to build.

The size of our aspirations equates to the size of the pool. If we have a large appetite with wild ambition, more effort is required. If you are content with less, your pool might look more like a bathtub. Neither is better or worse than the other. It's all about your own aspirations and desires.

Today, I'm proud to share that I get most of what I want from my work. I'm able to check a lot of boxes. It's not perfect, but it's far more than I ever hoped. I'm not done, far from it, but it's attainable. How long did it take to achieve? About a decade. Some may progress faster, which I hope they do, but I struggle to empathize with those who want it all on day one of their careers.

Work allows us to provide for ourselves, our families, and those we love. It can also be wildly purposeful and fulfilling. Over the arc of your career, I hope you taste the best parts of everything work can provide. My one world of caution is around timing. Setting too high of a standard too early can be costly. Oprah Winfrey is famously cited as saying, "You can have it all. Just not all at once."

I've spoken with a lot of engineers who don't find enough purpose, fulfillment, and enjoyment during their current season of work. I respond with a simple story of three workers laying bricks. Someone asks the first worker what they're doing. "Laying bricks, of course," the worker replies. Moving on to the second worker, the interrogator is met with, "Building a school." Finally, they ask the third worker who replies, "Housing the minds of our future." Each worker was performing the same task, but their perspectives found different meanings. We have the same choice in our own work as well.

28

GO BOLDLY FORWARD
It's your turn.

Until this point, we've worked to weave our (Tim and Morgan's) stories together with other voices from the industry. The goal was to make the information more digestible and the stories more cohesive.

Now is where our paths diverge. Despite the overlaps and shared insights, we were shocked when we started by how different our journeys began.

Tim's Beginning

Shortly after graduating college in 2005, I was sitting at my desk in the counseling center where I worked, wondering if this was what I was going to do for the rest of my life. While I enjoyed aspects of the job, there was so much of it that just didn't fit with what I wanted from life. On non-client days, I was able to get my work done before lunch and would spend the afternoons sitting around with nothing to do. I went to my boss and asked for more responsibility, but was told that there was no other work to be done and I should just enjoy the downtime. On those days, I decided to build my own website. At this time, WordPress was new and there weren't many options out there. So I decided to build the site myself, using only HTML and CSS. There weren't many online tutorials for learning how to code, so I walked a few blocks away to the library and checked out books on the topic. I would then spend these afternoons reading through the books and attempting to make sense of the content.

A few weeks in, I realized I needed help. I considered going back to school and went as far as applying to a Computer Science program. Time and money were not on my side. That's when I decided to look for a mentor. I searched for some local software agencies nearby and found a few. I emailed each of them, and offered to work for free or pay for mentoring. No one responded. I searched for months. One day, a friend connected me to Jack, who had also gone down the self-taught path. I reached out to Jack who informed me he was really busy, but if I wanted to, I could buy him dinner and he would answer my questions. This was the process of my learning over the next year. I would check out books from the library, spend downtime at work and evenings reading, and work through the material, all while saving my questions for the next dinner with Jack.

Being honest, I was disappointed with how hard it was to find a mentor. I was disappointed that none of the local shops would respond to my emails or even give me a chance. I should also point out that I was living in a fairly rural town, and while it might sound like excuses, back then, there just weren't as many resources to connect with other local developers. However, I stuck with it. The deeper I got into learning, the more I enjoyed it and realized this was something I could do long term.

About two years after I started, I met Kurt. He was in charge of the marketing department at a web development company in town, and they were hiring. It seemed like a long shot, but I knew I had to make an attempt. After chatting with him about the company, I applied and was shocked to be called in for an interview. As mentioned earlier in the book, I got the job. This was the first time I would be able to learn alongside other developers. I finally had my chance to talk with other folks about their coding styles instead of just reading about it in a book. My hard work had paid off. I was the happiest I had been so far in my career.

While Kurt quickly became a life mentor, he wasn't a software engineer. Almost two decades later, I've still never had a mentor specifically for software engineering. I've worked with great people who taught me different skills and helped me grow in my career, but I would have done anything for someone to come alongside me— someone to mentor me through my ups and downs. In Morgan's story you'll meet Jason. Someone like that would have been great but I've never been quite so lucky.

Morgan's Beginning

I admire Tim's story. Without much assistance, he is a testament to knowing what you want and not letting anything stand in your way, even if you're forced to trudge on alone.

I was more fortunate. Jason Ardell was my earliest mentor in software engineering. As he coached me through challenges, he would give me enough context to start, but leave room for me to thrash and discover things for myself. The moments when he would set me loose to figure things out were punctuated by one expression, "Go boldly forward."

Late nights, early mornings, and many Saturdays were spent shoulder to shoulder with Jason. As my experience grew, our relationship continued to evolve. I'd go longer stretches between asking for help, trying to piece things together alone. Fortunately, Jason was always there. I remember countless emails, text messages, and phone calls where he would talk things through. In many cases, the most helpful thing he would do is reply with a Google link that would give me just enough direction to help myself.

Since the beginning, I appreciated Jason's encouragement to "go boldly forward" because it articulated the tension between confidence, risk, and progress. It's about how you proceed as much as it's about the direction. At the core, it implies movement and courage.

Years later, my connection to *go boldly forward* has evolved. It's an attitude as much as a battle cry. I use it to challenge our team, friends, and myself. Similar to how "manifest destiny" was a cultural belief of 19th century Americans, "go boldly forward" summarizes how I hope to move throughout my career. It is woven throughout the work I do, the way I do it, and who I want to be—constant, courageous, and forward moving.

In our world of rapid change, those who stand still fall behind. Those who shrink back go unnoticed. My hope for myself, and those who choose to read this book, is to push for continual improvement, courageous action, and thoughtful discourse in order to advance the craft and the industry that supports it.

My push for boldness and courage is largely why I love working with code school students. The type of person who pursues unconventional education and operates far outside their comfort zone is someone I resonate with. I look for that attitude in people I hire, work with, and partner alongside. The freedom to experiment, challenge conventional thinking, and unearth new possibilities makes the process and the outcomes more enjoyable.

As we neared the end of our regular mentoring sessions, Jason left me with another compelling anecdote. As I looked back over the months of teaching, help, and encouragement he provided, I asked how I might repay him. Through all that time, a single dollar never changed hands. His response: "You don't owe me anything. Go do likewise." At the time, the thought of replicating his investment seemed daunting. Not only was I unsure where to start, I wasn't convinced I even understood what he meant.

With each passing year, I still work to honor his request. In pursuit of "doing likewise," I have responded to hundreds of emails from early stage engineers, written blog posts, launched apprenticeship programs for code school graduates, and created dozens of presentations I regularly deliver to code schools. His prompt, now almost a decade ago, was even the catalyst for this book.

Your Beginning

I'm inspired by how profoundly a single person, like Jason, can impact someone's life. It's equally as inspiring how the absence of such a person, in the case of Tim's story, can be just as impactful. "Your mileage may vary" is a popular tech phrase and our two experiences highlight the simplicity of that truth. There is no silver bullet or one-size fits all solution.

Now, it's your turn. If you've made it this far in the book, hopefully you have learned something. It's time to take action. As you do, I'd encourage you to find others to whom you might pass your learnings along the way.

There will always be risk, doubt, and uncertainty. May you too Go Boldly Forward.

29

LETTER FROM MORGAN
To whom much is given, much is required.

Dear Reader,

If you've read this far, thank you. It is the end, for now. Tim, myself, and our team worked hard to make it worth your effort. If you have feedback or ideas for how we can improve later editions, please reach out. We are still learning too.

When I was a kid, I heard the expression, "to whom much is given, much is required." It rang true then and continues to echo in my head, nearly two decades later. With each year, the weight of that responsibility intensifies. Sure, I've put in a lot of effort but I've also been given nearly every advantage in life. This book is one expression of how I might pay that forward and benefit others. If you found even just one or two nuggets useful, we consider it all worth it.

Whether email, LinkedIn, or somewhere else online, Tim and I are a default "yes" when code school students need help. I can't promise I'll have a job opportunity or magical advice but I respond to every message personally.

Time is your most valuable asset, thanks for trusting it to us.

Godspeed,

Morgan J Lopes

Endnotes

Preface

1. Bill Bryson, *A Walk in the Woods* (New York: Broadway Books, 1998).

Chapter 1: Engineers Solve Problems

1. Ryan Holiday, *The Obstacle is the Way: The Timeless Art of Turning Trials into Triumph* (New York: Portfolio/Penguin, 2014).

Chapter 2: Knowledge Work

1. Mihaly Csikszentmihalyi, *Flow: The Psychology of Optimal Experience* (New York: HarperCollins, 1990).

Chapter 3: Habits and Rituals

1. While many sources have come out about the 10,000-hours rule, our favorite is Malcolm Gladwell's book, *Outliers: The Story of Success* (New York: Little, Brown and Company, 2008).
2. Railscasts is a collection of screencasts and tutorials for Ruby on Rails from Ryan Bates. If you're not already using it, you can find it here: http://railscasts.com/.

Chapter 5: Imposter Syndrome

1. To read Harvard Business Review's entire article on Imposter Syndrome, see https://hbr.org/2008/05/overcoming-imposter-syndrome.

2. To hear Carol Dwek's entire TED Talk, "Growth Mindset", visit https://www.ted.com/speakers/carol_dweck.

Chapter 8: Learning by Doing

1. Anders Ericsson and Robert Pool, *Peak: Secrets from the New Science of Expertise* (New York: Houghton Mifflin Harcourt, 2016).

Chapter 14: Aligning with Employer Expectations

1. To read Google's entire study on Team Effectiveness, you can view it here: rework.withgoogle.com/guides/understanding-team-effectiveness/steps/introducton/.

Chapter 15: Hacking the Interview

1. Glassdoor republished this article from FINS from The Wall Street Journal on rules for following up. You can find it at: https://www.glassdoor.com/blog/rules-job-interview/.

Chapter 16: Planning a Successful Career

1. A helpful explanation of the Peter Principle can be found at the following Wiki link: https://en.wikipedia.org/wiki/Peter_principle.
2. Paul Graham writes about the Maker's and Manager's schedules on his blog here: http://www.paulgraham.com/makersschedule.html.

Chapter 20: Excellence, Not Perfection

1. To read Morgan's blog post, "Bringing Humanity Back to Email", go to: https://www.morganjlopes.com/bringing-humanity-back-to-email/

Chapter 22: Debate and Discourse

1. Sadly, this speech was at a private event, but you can take my word for it.

Chapter 23: Diversity, Inclusion, and Equity

1. Emily Chang, *Brotopia: Breaking Up the Boys' Club of Silicon Valley* (New York: Portfolio/Penguin, 2019).
2. Ijeoma Oluo, *So You Want to Talk About Race* (New York: Seal Press, 2018).
3. Carol Anderson, *White Rage: The Unspoken Truth of Our Racial Divide* (New York: Bloomsbury, 2016).
4. Nathalie Molina Niño and Sara Grace, *Leapfrog: The New Revolution for Women Entrepreneurs* (New York: Penguin Random House, 2018).
5. Austin Channing Brown, *I'm Still Here: Black Dignity in a World Made for Whiteness* (New York: Convergent Books, 2018).
6. Scott Page, *The Diversity Bonus: How Great Teams Pay Off in the Knowledge Economy* (Princeton, New Jersey: Princeton University Press, 2017).
7. To read more about Harvard Business Review workplace diversity study, visit: https://hbr.org/2017/03/teams-solve-problems-faster-when-theyre-more-cognitively-diverse.

Acknowledgments

First and foremost, we'd like to acknowledge our wives. Steph and Meg took a chance on us long before our careers were formed. Their patience and support continue to be a primary reason we are able to live our passions and interests. They endure all forms of scheming, ideation, and curiosity, for which we are grateful.

Next, we must shout out Emily Buchanan. This book project would not have happened without the steady pressure she applied. It must have seemed like herding cats. Not only did she bring order to the chaos, she graciously weathered our ignorance. Emily made sure our thoughts were discernible to newcomers and we made regular progress.

Over the course of this book and our experiences that shaped it, there have been numerous individuals who inspired us along the way. Many even directly contributed thoughts, feedback, and their own lived experiences to ensure this book provided a well-rounded perspective. We are forever grateful to the depth and breadth added by dozens of other voices. We do not have all the answers, but we are fortunate to have a strong network of others who care:

Alicia Barrett
Amelia Schulz
Annie Liew
Bethaney Wilkinson
Blanca Catalina Garcia
Brandon McLean
Brenton Strine
Brit Butler
Carlos Gonzalez
Charlton Cunningham
Cherie Lum

Chris Chancey
Clay Cribbs
Craig Rodrigues
Daniel McBrayer
DeeAnn Kendrick
Dominique Bashizi
Geraldine Galue
Glen Stovall
Imani Oliver
Jason Ardell
Jason Cambell
Jess Swart
Jessica Mitsch
Joey Womack
John Saddington
John Rae
Kathya Acuńa
Katie Bautista
Katie Watford
Kelly Leonard
Kevin Maldau
Kristina (Kristina Smith) Newton
Kurt Illian
Peter Barth
Lindsey Owings
Loren Norman
Martin Parets
Nerando Johnson
Nate Washington
Paige Niedringhaus
Princess Sampson
Riaz Virani
Richard Simms
Rose Lake
Sam Kapila
Shane Ardell
Stefanie M. Jewett
Terrance Jackson

Tobias Wright
Toby Ho
Toni Warren
Troy Wilson
Tun Khine
Usman Shahid
Vi Pham

Made in the USA
Monee, IL
28 August 2021

76710471R00105